TECHNOLOGY AND SCIENCE
IN THE INDUSTRIALIZING NATIONS

CONTROL OF NATURE

Series Editors
Morton L. Schagrin
State University of New York at Fredonia
Michael Ruse
Florida State University
Robert Hollinger
Iowa State University

*Controlling Human Heredity:
1865 to the Present*
Diane B. Paul

Einstein and Our World, second edition
David C. Cassidy

Evolutionary Theory and Victorian Culture
Martin Fichman

The Lysenko Effect: The Politics of Science
Nils Roll-Hansen

Newton and the Culture of Newtonianism
Betty Jo Teeter Dobbs and Margaret C. Jacob

*The Scientific Revolution:
Aspirations and Achievements, 1500–1700*
James R. Jacob

*Scientists and the Development of Nuclear Weapons:
From Fission to the Limited Test Ban Treaty, 1939–1963*
Lawrence Badash

Technology and Science in the Industrializing Nations, 1500–1914, second edition
Eric Dorn Brose

Totalitarian Science and Technology, second edition
Paul R. Josephson

Eric Dorn Brose

second edition

TECHNOLOGY AND SCIENCE
IN THE INDUSTRIALIZING NATIONS 1500–1914

CONTROL OF NATURE

Humanity Books

an imprint of Prometheus Books
59 John Glenn Drive, Amherst, New York 14228-2197

Published 2006 by Humanity Books, an imprint of Prometheus Books

Technology and Science in the Industrializing Nations, 1500–1914. Copyright © 2006 by Eric Brose. All rights reserved. No part of this publication may be reproduced, stored in a retrieval system, or transmitted in any form or by any means, digital, electronic, mechanical, photocopying, recording, or otherwise, or conveyed via the Internet or a Web site without prior written permission of the publisher, except in the case of brief quotations embodied in critical articles and reviews.

Inquiries should be addressed to
Humanity Books
59 John Glenn Drive
Amherst, New York 14228–2197
VOICE: 716–691–0133, ext. 207
FAX: 716–564–2711

10 09 08 07 06   5 4 3 2 1

Library of Congress Cataloging-in-Publication Data

Brose, Eric Dorn, 1948–
    Technology and science in the industrializing nations, 1500-1914 / [Eric Dorn Brose].— 2nd ed.
       p. cm.
    Includes bibliographical references and index.
    ISBN 1-59102-367-X (pbk. : alk. paper)
    1. Technology—Social aspects—History. 2. Technology—History. 3. Science—Social aspects—History. 4. Science—History. I. Title.

T14.5.B76 2005
609.4'0903—dc22

2005023252

Printed in the United States of America on acid-free paper

CONTENTS

Series Editors' Preface   vii

Preface   ix

Chapter 1. The Search for a New Atlantis   1

Chapter 2. Science and Technology in the World's Workshop   27

Chapter 3. Lands of Unlimited Possibilities   77

References   119

Index   129

SERIES EDITORS' PREFACE

This series of historical studies aims to enrich understanding of the role that science and technology have played in the history of Western civilization and culture, and through that in the emerging global civilization. Each author has written with students and general readers, not specialists, in mind. And the volumes have been written by scholars distinguished in the particular field. In this book, Eric Brose—a major interpreter of German history as well as the history of technology—draws upon his expertise to address one of the more extraordinary phenomena of Western history, the technological development of Europe in the nineteenth century. Out of it came unprecedented progress in life expectancy, urban infrastructure, transportation, and, not least, the art of war.

The aim of Eric Brose's book is to provide an overview of the key technological developments (among the many complex processes that scholars are currently exploring) that made for Western development and domination. Particular attention is paid to Continental European technological growth, but Britain is also present in his account. In addition, the volume engages with the questions and debates that inform contemporary scholarship. It seeks to make the technology and science accessible and readable by anyone.

The current debates and overall perspective in this volume, and the series to which it belongs, emphasize the "Control of Nature." While not excluding a discussion of how knowledge itself develops, how it is constructed through the interplay of research into nature with the values and beliefs of the researcher, the series looks primarily at how science and technology interact with economic and political life, in ways that transform the relationship between human beings and nature. In every volume in this series, we are asking the student to think about how the modern world came to be invented, a world where the call for progress and the need to respect humanity and the environment produce a tension, on the one hand liberating, on the other threatening to overwhelm human resources and ingenuity. The engineers and statesmen whom you will meet here, although excited by the revolution in modes of production they were producing, could not have foreseen the kind of power that early-twenty-first-century science and technology both promise and deliver. But they were also dreamers and doers—even shrewd promoters—who changed forever the way people experience the natural world.

Margaret C. Jacob
Rosemary Stevens
Spencer R. Weart

PREFACE

This book surveys the history of technology and science over four centuries. It begins with the opening of the modern historical epoch around 1500 and ends with the outbreak of World War One in 1914.

One goal of the study is to weave various historiographical threads—separate or only partly woven until now—into one fabric. The history of technology, the history of science, and the history of economic development leading to the Industrial Revolution have developed to a very great degree as three histories. For decades historians of science and technology agreed with one another that until the 1900s there was little causative interaction between the subjects they studied. Economists treated science and technology as "residual" factors that were important only to the extent that they could be measured. Busy pursuing detailed monographic research in these fields, moreover, few historians attempted to demonstrate the relationship between science, technology, economy, and general developments.

A. E. Musson and Eric Robinson (1969) made a start toward integration by demonstrating that science had indeed contributed to the technology of the English Industrial Revolution. Margaret Jacob (1981, 1988, 1997) furthered this historiographical revolution with her nuanced studies of the connection among culture, science, and technological change in England and western Europe during the seventeenth and eighteenth centuries. Larry Stewart (1992) added to these insights with his analysis of the rise of public science in eighteenth-century England. A few years before this, Bertrand Gille (1986) wrote a comprehensive history of technology that devoted considerable attention to European developments after 1500. He also advanced the thesis of "technological systems" that either evolve or deteriorate as societal demands change (see chap. 1). In explaining technological change, Gille appreciated economic and scientific factors as well as the role of social and political institutions. He was the first to admit that his work was only a sketch, however, and, perhaps because of this, it has not received the attention that it deserves. The present study benefits from these newer works in the history of science and technology, while also drawing on a whole series of recent monographs by economic historians such as Maxine Berg (1994) and Joel Mokyr (1999). It is a work of synthesis—an "interpretive" survey.

The author and publishers hope that this book will interest scholars and graduate students of history. A more important goal, however, is to present the arguments and narration in a fashion that engages beginning undergraduate students. For it is high time that we break definitively from the approach in introductory courses that pushes science, technology, and other newer fields (e.g., gender studies) to the upper level as elective or so-called special topics courses, thus preserving the traditional view of "real" history as past politics. We hope the subject matter of this book, in other words, will be moved into the mainstream.

Philadelphia

September 2005

I

THE SEARCH FOR A NEW ATLANTIS

The English travelers had been lost in the South Seas for an endless succession of days before they sighted land. As hour by hour the narrow strip on the horizon grew larger and greener, the sailors' mood brightened to euphoria.

Soon they could discern the outlines of a remarkable city with strong walls. Before the hungry men could enter port, they were met by a small boat. Its leader, showing no signs of fear or distrust, presented the confused foreigners with a shiny scroll written in ancient Hebrew, Greek, and Latin: "Land ye not, none of you. Provide to be gone from this coast within sixteen days, except you have further time given you." The Englishmen appealed for mercy and, after three hours, were given leave to come ashore. They were escorted to Strangers' House as guests of the island kingdom of Bensalem.

After three days, the Governor of Strangers' House visited their quarters. They listened intently as he told them about a wonderful civilization, many thousands of years old that predated European antiquity and had continued to develop, independently of Europe, and unknown to it, since the end of Roman times. The most important turning point in the history of Bensalem came in the reign of Salamona, many centuries before the fall of Rome. This wise and noble king had decreed the establishment of a scientific order "dedicated to the study of the works and creatures of God." The College of the Six Days Works, later known simply as Salomon's House, had evolved over nineteen hundred years into a true blessing for the people who were now the beneficiaries of advanced arts and sciences. The Englishmen were told of the impending visit of one of the thirty-six elders of the House of Salomon. "I will provide you and your fellows of good standing to see his entry."

After observing the regal procession of the great scientist and his entire retinue, the visitors were granted an audience with him. Clothed in a long robe of fine black cloth, with wide sleeves and a cape, the great one sat on a richly adorned throne as they entered. His head was covered with an embroidered cloth of state made of blue satin. "God bless thee, my sons. I will give thee the greatest jewel I have. For I will impart unto thee, for the love of God and men, a relation of the true state of Salomon's House."

The high priest of knowledge proceeded to describe the amazing institution and its many accomplishments. There were deep caves and high towers for investigating earth, minerals, weather, and sky; huge chambers for the simulation of rain, snow, thunder, and lightning; parks and laboratories for raising and crossbreeding plants, fish, and animals; experimental gardens and kitchens for preparing the most nourishing and exotic foods; medical institutes for testing herbs and potions to cure disease and prolong life; special microscopes, telescopes, and hearing aids; devices for flight and underwater travel; fiery furnaces, powerful engines, and ingenious machines for producing all manner of goods; and many other marvelous inventions. The main purpose of so much research, he stated, "is the knowledge of causes, the secret motions of things, and the enlarging of the bounds of human empire [for] the effecting of all things possible."

Indeed there was an elaborate division of labor among the fathers of Salomon House: some, like hermits, engaged in pure research; some compiled and categorized the results; some considered "how to draw out of [the experiments] things of use and practice for man's life"; others pondered new experiments based on the old; while still other elders, called "Interpreters of Nature," expanded experimental discoveries "into greater observations, axioms, and aphorisms."

The Englishmen listened and marveled—and wished that England could be so (Bacon 1955b, 546, 563, 572, 582).

AN EARLY MODERN TECHNOLOGICAL SYSTEM

Francis Bacon wrote *The New Atlantis* shortly before his death in 1626. He was an impressive combination of politician, lawyer, philosopher, moralist, and scientist—in short, a true Renaissance Man. For centuries, historians eulogized Bacon as one of the founding fathers of the seventeenth-century intellectual revolution that gave birth to modern science. More recent findings position him somewhere in the midst of three traditions of early modern science, or "natural philosophy," all of which mixed the archaic and the bizarre with notions that strike us as more modern. There was the dominant "organicist" school in science, grounded in Aristotle, which explained nature in biological terms like growth and decay; the "magical" tradition, dating back to Plato, which believed there were secret, animate powers at work behind nature's mysteries; and the "mechanical" approach to science, based on the works of Archimedes, which saw the natural world as an inanimate mechanism. Bacon seems to have been influenced by all three traditions, using human analogies like "the senses" to describe air, for instance; at other times he referred to "the secrets of stones" or claimed that the study of mechanics would have "the most radical and fundamental" effect on natural philosophy. But it was his irrepressible belief in progress and constant urging to surpass the ancients that preserve Bacon's reputation as "a modern." How long, he complained, would men let the Greeks and Romans "stand up like Hercules' Pillars, beyond which there can be no sailing or discovery!" It

4   Technology and Science in the Industrializing Nations, 1500–1914

is no coincidence, therefore, that utopian Bensalem was presented as a unique civilization that had outdistanced ancient Europe. The title page of another work, *The Great Instauration* (1620), depicted ships that were about to sail past the Pillars of Hercules on a voyage to new intellectual worlds (see figure 1.1) (Jones 1965, 43–44; Kearney 1986, 94; Webster 1976, 279).

Bacon's sanguine, imaginative outlook on the future was certainly a product of his faith in science. It also sprang from his belief that scientists could learn from the step-by-step methods and steady advances being made by early modern craftsmen and engineers. A systematic consideration of their work would not only facilitate the spread of useful technology but also give natural philosophers "a more true and real illumination concerning causes and axioms." Bacon rejected the prejudice which led many of his contemporaries to deem it "a kind of dishonour unto learning to descend to inquiry or meditation upon matters mechanical." For such "supercilious arrogancy" flew in the face of the many impressive breakthroughs made in recent times by Europe's most skilled artisans (Bacon 1955a, 232–30).

In the area of shipbuilding and navigation, for instance, the ambitious voyages of discovery were the moon shots of that era. Spanish and Portuguese expeditions had reached the Americas and India in the 1490s, and an English ship circumnavigated the globe in the 1570s. Preparing the way for these long, deepwater journeys was an ingenious combination of techniques mastered by the Portuguese in the course of the fifteenth century: broad, sturdy hulls; multiple decks and masts; the triangular lateen sail of the Arabs (which afforded maneuverability in headwinds and crosswinds); European square rigging (for rapid windward progress and coming about); and the sternpost rudder. Navigators used crude compasses, simple astrolabes, or quadrants and observed the polestar or the midday declination of the sun to determine latitude. No solution for longitude emerged for another three hundred years, but the basic technology of ocean travel was set until the dramatic changes of the mid-nineteenth century involving steam technology (Hall 1967; Parry 1966; Schmidtchen 1992).

Figure 1.1: Title page to Francis Bacon's *Instauratio Magna* (1620).

European expansion and discovery had positive effects on agriculture and horticulture. From the Americas came potatoes and corn; from Africa, cloves and cinnamon; from the Near East, artichokes, melons, and cauliflower. In Italian botanical gardens experiments with carrots, lettuce, and strawberries were underway that would have earned nods of approval from the fictional New Atlantians. These developments coincided with a rise in population and prices after 1500, moreover, that stimulated land reclamation projects to expand production. The Pontine marshes of central Italy and the hinterlands of Venice, the fenlands of East Anglia and seaside "polderlands" of Holland, as well as new farmland in eastern Germany and the Baltic were brought under the plow. Simultaneously, a determined search for new agricultural techniques registered some success, especially in Flanders as innovators began to rotate crops. But a revolution in European agriculture remained largely the subject of experts' dreams in the sixteenth century. Despite the rapidly escalating number of printed treatises about farming, famine revisited Europe with a dreary regularity. Agricultural proficiency was one of the biggest gaps in Europe's technological system (Gille 1966; 1986, vol. 1; Hall 1967).

The plethora of sixteenth-century farming manuals resulted from another important technological change of the era preceding Francis Bacon: the printing press. Like Portuguese ships of the fifteenth century, the printing press combined a cluster of known techniques that were improved upon and employed in innovative ways. Probably first used in the 1440s by Johann Gutenberg of Mainz, Germany, it integrated the technique of engraved stamps used by goldsmiths; the tin-lead alloys used by pewterers for these engraved stamps, called "type"; a thick, linseed oil–based ink that painters knew would not run on metal; the linen press, found in every well-to-do household; and paper, which had made its way to Europe from the Far East. By casting separate, movable metal types of accurate dimensions that held firmly together during printing and could be easily reassembled, Gutenberg could reproduce page after page with relative ease (see figure 1.2). By 1500 there were more than a thousand printing presses in Europe, most in Germany and Italy, which had produced about thirty-five thousand editions. The accompanying impact on the manufacture of paper in mills (which consumed a growing amount of wood) enhanced the revolutionary nature of this technology, but, like telephones, computers, and e-mail in our day, it was the increased volume and velocity of exchanged knowledge that had the most profound effect on early modern Europe (Gille 1986, vol. 1; Hall 1967; Pacey 1980; Schmidtchen 1992).

Other noteworthy technological innovations emerged during the 1400s and 1500s. The most important of these were labor-saving devices that reflected a shortage of hands still lingering from the Black Death of the 1300s. One of the simplest but most widespread breakthroughs came with the spinning wheel. A "flyer" attached to the older spindle wheel combined the winding and twisting of yarn into one motion controlled via a crank-and-treadle system by the operator's foot (see figure 1.3). Other textile machines were more complex, like the silk and woolen looms that began to appear or an English mechanism for knitting stockings. Windmills had developed into modern form by the late 1500s, furthermore, mainly in England and coastal northern Europe, where they ran grindstones and sawmills with as much as fifty horsepower (see figure 1.4). "The fully developed windmill," writes A. Rupert Hall, "was the most elaborate large-scale machine before the steam engine, and the millwright was the master mechanic of his age." Waterwheels generated even more power, evolving quickly in numbers and scale from the cruder devices of earlier times. Used in mines, foundries, grain mills, and sawmills as well as for hoists, cranes, lathes, large hammers, and textile machinery, these heavy timber wheels and shafts appeared along every important river and stream of early modern Europe. Already by 1550 there were forty different manufacturing processes in Europe that utilized this technology. Mechanization, in turn, spawned a new generation of hand tools. Improved jigsaws, hacksaws, planes, molders, drills, and pliers enabled skilled craftsmen to construct the machines demanded by an expanding society (Gille 1986, 1:83).

These unfolding technological feats—particularly the most impressive breakthroughs in navigation, printing, and mechaniza-

8   Technology and Science in the Industrializing Nations, 1500–1914

Figure 1.2: A sixteenth-century printing shop. From Charles Singer, ed., vol. 3 of *A History of Technology* (1957), p. 395.

Figure 1.3: Dutch (or Irish) spinning wheel with flyer. From Charles Singer, ed., vol. 3 of *A History of Technology* (1957), pp. 160–61.

Figure 1.4: Dutch windmill showing a high degree of mechanization.
From Charles Singer, ed., vol. 3 of *A History of Technology* (1957), p. 91.

tion—convinced some Europeans that a golden age was dawning. This conviction was steeped in a highly emotional and zealous creed known as "millenarianism," the belief that a thousand-year rule of the saints with God was beginning. This era of peace, prosperity, and expanding knowledge—the Millennium—would be followed by the end of time and Judgment Day. Bacon's *New Atlantis* was clearly part of this millenarian tradition. Bensalem was a walled city with powerful military technology—we are told of "ordnance and instruments of war exceeding your greatest cannons and basilisks" (Bacon 1955b, 581)—but it was a devout and confident civilization intent on peaceful scientific endeavors. William Shakespeare's *The Tempest* (1613) expressed the same yearning for a haven from the world's storms and a future of peace and justice controlled by magical scientists like Prospero. But other Europeans, frightened by a seemingly endless succession of increasingly violent wars, interpreted the Millennium as a historical phenomenon that began in late Roman times and was yielding now to an apocalyptic and catastrophic end of the world. Albrecht Dürer's early-sixteenth-century woodcut depicting the Four Horsemen of the Apocalypse—war, fire, pestilence, and famine—trampling down the high-born and the low became the most famous expression of these deep-seated fears (see figure 1.5) (Firth 1979; Jacob 1988; Kearney 1986).

It was understandable that early modern Europeans either feared the Apocalypse or wished desperately for a respite in the Millennium. Indeed few eras in history were as torn asunder by military conflagrations as this one. The Italians Wars (1494–1559), the European Wars of Religion (1562–1598), and the Thirty Years War (1618–1648) wreaked havoc on successive generations, as did civil wars in France (1562–1598), Sweden (1592–1600), Russia (1598–1613), and England (1642–1649). But casting a pall over the entire period was the threat of Ottoman Turkey. First felt in 1453 with the fall of Constantinople, the peril drew closer in 1521–22 with the capture of Belgrade and Rhodes and did not pass until the lifting of the yearlong siege of Vienna in 1683. These conflicts play an important part in our story, for they stimulated technological change (Bonney 1991).

Figure 1.5: Albrecht Dürer's *Apocalypse*.

The French army that invaded Italy in 1494 brought along large numbers of mobile siege guns. Most cities wisely surrendered after token resistance, but in a mere eight hours the new cannons reduced one fortification near Naples to rubble. It had recently withstood a siege of seven years. French founders had substituted lighter bronze (and to a lesser extent brass) artillery for the huge wrought-iron "bombards" of the early 1400s (see figure 1.6), thus enabling armies to transport many cannon rather than make a few of them on the spot. Moreover, the French guns fired cast-iron balls in place of the awkward, less destructive, and expensively shaped stone projectiles. The result was a frightening revolution in artillery that spread to the other armies of Europe and Turkey during the early sixteenth century.

After the 1540s, English guns made of cast iron offered a much cheaper alternative to the qualitatively superior bronze variety cast mainly in Germany. The latter began to experience stiffer competition in the 1620s, however, from Swedish cast-iron cannon of much higher quality. The new demand for cast-iron cannonballs raised European iron production to sixty thousand tons in 1500—a figure that soared to over one hundred fifty thousand tons before 1600. The production of bronze and brass rose in similar proportions as the terrible wars continued (McNeil 1982; Pacey 1980; Schmidtchen 1992; Troitzsch 1991).

As occurs so often in history, war and technological change were inextricably linked. The making of bronze—an alloy often consisting of eight parts copper and one part tin—did not change dramatically, for the copper smelters and bell founders who cast the first cannons had already developed their furnace art and molding techniques to an advanced state by the 1400s. Rather, iron making changed the most. Prior to the late fifteenth century, iron ore was roasted in an open hearth, then purified by a process of repeated heating and hammering to remove extraneous matter. Responding to the new military demand, ironmongers in Liege began to melt iron ore in tall (twenty-five-foot) furnaces fueled with charcoal and brought to red heat with bellows driven by waterwheels. The molten iron was tapped from the blast furnace out of a bottom opening into molds, or "pigs." The resulting crude or "pig" iron could be remelted

Figure 1.6: Progression (*top to bottom*) of cannons in the fifteenth century. From Berhard Rathgen, *Das Geschütz im Mittelalter* (1828), Table 4/12; A. Essenwein, *Quellen zur Geschichte der Feuerwaffen* (1877) 2: pl. A 21–22; *Königliche Hof- und Staatsbibliothek* (Munich).

and cast into balls or cannon, or further refined to reduce carbon content. This more malleable finished product, known as wrought iron, could be hammered by blacksmiths into a wide variety of implements for town and country. The new iron-making methods spread throughout France, England, Germany, and Scandinavia in the sixteenth and early seventeenth centuries as entrepreneurs, eager to profit from their monarchs, learned the art quickly (Gille 1986, vol. 1; Smith 1967b).

Europe's succession of wars also affected mining. Responding to the greater demand, miners extracted copper, tin, and iron ores in much greater quantities. But there were other, less obvious changes. Thus silver mining in central Europe took off as expanding trade increased the need for coinage—and as financially strapped kings struggled to pay for the new weapons of destruction. In the Freiberg district of Saxony, for instance, silver production shot up from sixteen thousand ounces annually in the 1400s to a hundred fifty thousand ounces by the mid-1500s. In 1530, nearly a hundred thousand men were employed in mining throughout the Holy Roman Empire of Germany. All Europe produced around a million ounces of silver by 1600. Moreover, as ship construction and paper and metal production depleted wood supplies, coal became a tempting substitute fuel, especially in England, where trees became scarcer during this period. Accordingly, British coal-mine output rose meteorically from two hundred thousand tons a year in the 1550s to about three million tons in the 1680s (Nef 1957; Pacey 1980).

Mining technology blossomed as a result. The ancients had mined in pits or larger quarries that followed a mineral vein from the surface into the earth. Early modern Europeans devised and perfected a method known as tunnel mining, whereby ore or coal was extracted from a ridge or hillside by burrowing upward from the valley floor. Tunnels ascended at three- to twelve-degree angles and were shored every few meters with wooden timbers. The downward exit slope facilitated water drainage and the removal of minerals. Miners worked minerals using either an "underhand stoping" method—hacking in a steplike pattern beneath the haulage tunnel, then shoveling ore upward from platform to platform—or "overhand stoping," where one worked in the reverse fashion above the tunnel. As the demand for metals grew, however, working areas descended farther and farther below the haulage tunnel. By the mid-1500s, in fact, the busiest mines in Saxony and Bohemia anticipated modern times by sinking deep vertical shafts, then tunneling horizontally into the veins. Approaching depths of six hundred feet, central European mines had to devise methods for removing ore and water. One of the most famous treatises on mining, *De re metallica* (1556) by Agricola (Georg Bauer), depicts gargantuan waterwheels that powered elevators, ore crushers, and pumps through an ingenious system of levers, gears, and connecting rods (see figure 1.7) (Gille 1986, vol. 1; Pacey 1908; Smith 1967b; Troitzsch 1991; Usher 1957).

The expanding output of metals encouraged and facilitated new uses for them. And these factors, in turn, spurred even more production. Metal fittings onboard ships; machine parts in the workshop, mill, and mine; type for printing presses; plows and other tools on the farm; and a host of items in the home began to be fashioned from iron, steel, pewter, brass, and bronze. "We could even say," writes historian Bertrand Gille (1986, 1:544), "that the metal age really began during this epoch."

Gille sees the spreading use of metal, in fact, as part of a wider technical phenomenon. In his magnum opus, *The History of Techniques* (1978), this insightful French scholar advanced the thesis that early modern Europe was evolving a new "technological system." The purpose of such a system in this—and indeed every—epoch was, simply enough, to meet the material demands of society. The basic component parts of a technological system are those that perform a technical act or job. Gille refers to the simplest of these—tools and levers, for example, or uncomplicated machines such as the spinning wheel—as "technical structures." But some jobs require a more complex grouping of separate techniques. Deepwater sailing, printing, or blast furnaces, for instance, are "technical ensembles." Vertical chains of technical ensembles that together produce a finished product he labels "technical concatenations." Thus the manufacture of clothing linked the separate job ensembles of spinning flax, wool, or cotton into thread; weaving thread into cloth; and

Figure 1.7: Use of water power in central European mines.
From Bertrand Gille, vol. 1 of *The History of Techniques* (1986), p. 538.

bleaching and dyeing the final fabric. Similarly, metals moved along the chain of ensembles from mining to smelting, finishing, and cutting and machine-tooling. Lending a certain coherence to the various technologies of a system were linking technical structures and materials. In Early Modern Europe these were, first, the crank and connecting rod for converting rotary to reciprocal motion, which appeared in spinning wheels, windmills, and waterwheels, and second, wooden structures increasingly reinforced with (and sometimes replaced by) metal (Gille 1986, vol. 1).

The technological system of early modern Europe was an impressive response to society's needs. Ultimately, however, its potential proved to be very limited. First, as noted above, Europe failed to solve its agricultural problems. Second, in order to continue to develop, mechanization required ever greater sources of power and energy. Wind had the obvious drawback of restricting mills to windswept coastal areas, and waterpower, too, limited production sites to rivers. The waterwheel was also unreliable during harsh winters or droughts and generated no more than ten to twenty horsepower in the early "undershot" designs (i.e., where water hit the bottom). As witnessed by the wooden mining monstrosities, moreover, wood was a relatively weak material that compensated by pushing the scale of construction to the outer bounds. Iron machines or machine parts offered one solution to this constraining problem, but the use of charcoal as a ferrous metallurgical fuel pushed up costs as iron mongers exhausted trees in the immediate vicinity of their furnaces and foundries. As explained earlier, this was a more pressing problem in England, where attempts to substitute coal for charcoal failed to produce an iron free of impurities. Finally, the military origins of much of Europe's technology began to undermine the system itself as wars escalated in frequency, violence, and destruction. During the Thirty Years War, for instance, armies disrupted trade, sacked city after city, and ruined many of central Europe's most productive mining and metallurgical centers.

As impressed as early-seventeenth-century thinkers like Francis Bacon were with the accomplishments of handicraftsmen and engineers, therefore, it was clear that Europe had to pass through a long "millennium" of basic scientific research and theoretical inquiry before solutions to technological problems produced a New Atlantis.

THE INTERACTION OF SCIENCE AND TECHNOLOGY

What can we say, then, about the relationship between science and technology in the early modern era? By the mid-fifteenth century, some technicians (i.e., artisans and engineers) had advanced beyond the commonsense approach to problem solving (which is sometimes mistaken by historians for an experimental scientific method) to a more sophisticated classification of problems and notions, as well as a curiosity about general laws and concepts. The great Italian artist-engineer Leonardo da Vinci (1452–1519) was undoubtedly the best example of this phenomenon. According to Gille,

> A science with a liking for reality, ready to check its results against experiment and a [technology] anxious to obtain explanations that would be more general and of greater validity, and which, moreover, had become increasingly able to make its own calculations, must inescapably have had close contacts with each other. And as soon as the one abandoned some of its abstraction and the other looked for generalization, an encounter became inevitable (1966, 220).

Although it would be facile and inaccurate to assert that science at this time was rooted in the technicians' search for solutions to vexing problems, it seems certain that science followed technology's lead in a vast majority of important cases. Emphasis on the cross-fertilization between artisans and early modern scientists—the so-called scholar and craftsman thesis—remains an important school in the historiography of the scientific revolution (Rossi 1970, 2000).

Mathematics is a good example of these kinds of interactions. Most natural philosophers of the fifteenth and sixteenth centuries adhered to an Aristotelian science that had little need for numbers. Mathematics, in fact, was widely held to be the preserve of lowly technicians or clerks. Accordingly, the first printed arithmetic manuals were written for young people interested in commercial careers. Algebra developed in the same practical milieu, designed primarily to explain premiums, rebates, and other problems of double-entry bookkeeping. Regiomontanus's early work on trigonometry, published in 1533, responded essentially to questions of measurement and surveying, while Simon Stevin, a late-sixteenth-century Dutch mathematician, turned to practical problems encountered with engineering dikes, water mills, and fortifications. He also devised tables for calculating interest. Even Desargues, the great French geometrician of the seventeenth century, was initially concerned with practical applications of geometry to painting and architecture (Rossi 1970; Gille 1966).

Physics, in particular, the science of dynamics, is perhaps the best example of our case. The modern study of the relationship between motion and the forces affecting motion grew essentially out of ballistics. By the late 1400s gunnery presented artillerymen with the difficult challenge of striking the top of a castle or city wall from a great distance. The explosive charge, the angle of the cannon, and the material, size, and speed of the projectile all seemed to affect accuracy and the force of impact. Moving beyond battlefield trial and error, the French soldiers who invaded Italy in 1494 fired rounds at canvases stretched between poles on a beach to learn about range and trajectory. Leonardo da Vinci, an able technician, studied cannon trajectories and force of impact even more systematically, as did Niccolo Tartaglia (1500–1557), an Italian engineer cum mathematician who experimented with cannon elevation and range. It is interesting that Tartaglia, an Archimedian enthusiast who translated the works of the great ancient mechanist in 1544, nevertheless offered no challenge to the Aristotelian notion that moving bodies had "impetus," a quality "possessed" by a projectile much as a plant had life. The first really scientific study of dynamics awaited Galileo Galilei's mathematical description in 1638 of a projectile's parabolic trajectory and other experiments that definitively broke with the older concept of impetus (see below) (Gille 1966; A. R. Hall 1957).

Mathematics and physics grew more closely related during the sixteenth century as both sciences evolved in opposition to Aristotelian dogma. By the early 1600s Bacon could observe that "if physics be daily improving and drawing out new axioms, it will continually be wanting fresh assistance from mathematics." Bacon did not always agree with his Italian contemporary, but the reference here was undoubtedly to the great "mechanical" scientist Galileo Galilei. A professor of science at the University of Padua from 1592 to 1610, Galileo pictured himself the Archimedes of his era. Like the ancient defender of Syracuse, he was drawn to problems of dynamics and statics. The raw material for his projects was personal observations in the arsenals and shipyards of the Venetian Republic that employed him.

In one famous study, Galileo hypothesized mathematically, then demonstrated experimentally, that the speed of a body freely falling from rest was the square of the time required to cover the distance. He used this to show that objects accelerated in a free fall, but not after leaving the muzzle of a cannon, when they moved at a constant speed. There was no more talk of "impetus." In another study, Galileo sought to explain the shipyard wisdom that large ships were weaker in proportion to their size than smaller vessels. He demonstrated the principles at work with an analysis of beams, arguing that the maximum supporting force of a beam embedded in a wall equaled the weight of the load plus the weight of the beam itself. Because, with larger beams, the weight increased more than the supporting force, a point would eventually be reached when the beam could support only itself, and thus no load. As Arnold Pacey argues (1980, 114), the theory had wide application in mechanics, for "in principle [it] could be extended to cover the working of every type of machine—water-wheels, horse-gins, cranes, and so on—and the equilibrium of every type of structure—ships, buildings, animals' skeletons or the framing of machines." Knowledge and application of these principles in workshops and mines, of course, was an excruciatingly slow process that depended, as we shall see in chapter 2, on a variety of social, political, and cultural factors (Butterfield 1957, 118; Kearney 1986).

When turning to metallurgy, we find a similar relationship

between science and technology—with the former curious about, and essentially following, the latter. Since the earliest ancient times, working with metals was a process that had advanced empirically from on-site experience. The first metallurgical treatises printed after 1500 performed a valuable service by circulating information about these accumulated techniques. Thus Vanuccio Biringuccio (1480–1539) noted the importance of steel's color in the quenching process: "Because the first color shown by steel when it is quenched while fiery is white, it is called silver; the second which is yellow like gold is called gold; the third which is bluish and purple they call violet; the fourth is ashen gray. You quench them at the proper stage of these colors as you wish them more or less hard in temper." Similarly, an early-seventeenth-century manual described the importance of observing the texture of bronze: "Bell founders judge the quantity of tin they should put in bell metal by breaking a piece of the material before they cast it . . . because if they find the grain too large they put in more tin, and if it is too fine they augment the copper." The most important contribution of "science" during much of the early modern period, odd as it may seem, came from the alchemists' quest for a magical process to convert common metals into gold, for they advanced furnace technique and practical chemical knowledge. By the seventeenth century, however, scientists of a more modern stripe were aiding metallurgists by examining metal texture under microscopes, measuring the densities of metals, and testing metal strength and elasticity (Forbes and Smith 1957, 3:36, 57).

While science was usually the handmaiden of technology at this time, it is important to note that this was not exclusively the case. The search for an adequate heat engine is a very important example of information flowing in the opposite direction. The story begins in England, where coal-mine owners experienced severe drainage problems in the late 1500s as shafts sank deeper to facilitate expanded production of this mineral fuel (see above). Expensive waterwheel and horse-gin designs imported from central Europe failed to meet the challenge in a cost-effective manner because profit margins for coal were lower than those for precious metals mined in Germany. With coal output soaring in the 1610s and 1620s, David Ramsay, an ambitious inventor, attempted to pump water from the mines utilizing ideas attributed to Hero of Alexandria, a tinkerer from ancient times. Giambattista della Porta, a neo-Platonist from Naples fascinated with "natural magic," had published a book in 1606 that described Hero's fired pneumatic gadgets. But Ramsay could not make the machines work (Nef 1957; Pacey 1980; Troitzsch 1991).

Meanwhile, scientists who were intellectually curious about Hero's devices made one significant discovery after another. Gasparo Berti, a Roman mathematician, discovered atmospheric pressure in 1642, and Otto von Guericke of Magdeburg, Germany, produced the first vacuum in 1654. Guericke took his experiments one step further in 1661 by creating a vacuum in a cylinder rigged with a piston and pulleys. The vacuum drew the piston downward, lifting heavy weights on the other side. It was left to a French Huguenot, Denis Papin, to experiment in 1673 with the idea of alternately heating water and condensing steam in a cylinder to move a piston. It would seem that Thomas Newcomen learned of Papin's work sometime in the 1690s—information that the Englishman used over two decades of trial and error to build the first really useful steam engine based on the principle of alternating or reciprocating motion. The device was installed in a coal mine near Wolverhampton in 1712. As we shall see in chapter 2, complex "technical structures" such as this would create the basis for a new technological system in the late eighteenth century (Allen 1977; Pacey 1980; Troitzsch 1991).

Thus science could make important contributions. That it should do so in a systematic, institutionalized manner was Francis Bacon's dream, in fact, when he drafted *The New Atlantis*. Indeed Salomon's House was to be the model for a state-run scientific college that would usher in the new millennium. The first such research institute in Europe, the short-lived Academy of Experiment, was founded in Florence in 1657. The example experienced more success with the Royal Society of London in 1662 and the French Academy of Science in 1666. Both sponsored scientific meetings, published papers and transactions, and undertook a whole range of experiments in practical areas like navigation, gunnery, fortification, agriculture, energy, and manufacturing technology. But, as Dr. Samuel

Johnson, a widely respected man of letters, observed nearly a century later, Bacon's utopia was slow in coming:

> When the philosophers of the seventeenth century were first congregated into the Royal Society we are told that great expectations were raised of the sudden progress of the useful arts. The time was supposed to be near when engines should turn by a perpetual motion, and health be secured by the universal medicine; when learning should be facilitated by a real character and commerce extended by ships which could reach their ports in defiance of the tempests. But that time never came. The Society met, and departed, without any visible diminution of the miseries of life. The gout and stone were still painful; the ground that was not ploughed brought forth no harvest, and neither oranges nor grapes could grow upon the hawthorn.

There were notable exceptions, of course, in both England and France. Thus Papin's breakthroughs in steam technology were conducted with Dutch-born Christiaan Huygens under the auspices of the French Academy of Science that he had helped to found. But Johnson was basically correct: the royal societies offered no scientific solution to a technological system that was stretched beyond its limits (Armytage 1965; Merton 1970; Pacey 1980).

In the long run, however, technological change would receive great impetus from the one classic example of the scientific revolution that we have thus far overlooked. Indeed by turning their gaze to the heavens in what we would today call "pure" scientific research, early modern scientists were destroying what remained of Aristotelian concepts. With practical technological problems of navigation forming only a pale celestial backdrop, neo-Platonists like Nicholas Copernicus (1473–1543), Tycho Brahe (1546–1601), and Johannes Kepler (1571–1630), and Archimedian champions like Galileo Galilei (1564–1642), waged a nasty feud against a conservative "organicist" establishment that attached religious significance to notions of a fixed and motionless earth surrounded by sun and planets moving in circular orbits at uniform speeds. The first salvo came with Copernicus's assertion in 1543 that the earth revolved around the sun, not vice versa. Another breakthrough occurred in 1609 when Kepler argued that the planets' orbits were elliptical, not circular, and that they accelerated when closer to the sun. The destruction of the antiquated concept of motion in the universe, and the substitution of a new scientific synthesis based on the law of gravity, was completed with the publication in 1687 of Sir Isaac Newton's famous *Principia*. It would be remiss not to point out that the great work appeared under the imprimatur of the Royal Society of London. As we shall see, the scientific and intellectual stage was now set for England's Industrial Revolution (Gribbin 2002; Shapin 1996; Jacob 1988).

II

SCIENCE AND TECHNOLOGY IN THE WORLD'S WORKSHOP

The highlands of northern England silhouetted the provincial town of York on a bleak spring day in 1772. A heavily laden wagon turned away from the river and rumbled slowly into the city, stopping in front of a book dealer's storefront. Soon a crowd of puzzled townspeople gathered around the wagon, wondering out loud about its mysterious contents. One by one, the driver unloaded the strange-looking devices and moved them into a large front room. There were metal globes and spheres, pumps, magnets, microscopes and telescopes, miniature cranes and pile drivers, as well as working models of windmills, waterwheels, bucket engines, steam engines, and a canal and colliery. When all the contraptions were nicely arranged, as if on display, the driver nailed a notice to the doorjamb.

The crowd moved closer to see what it said: "Announcing a Course of Lectures on Natural and Experimental Philosophy. By A. Walker. The lectures require one guinea a gentleman's transferable ticket, half a guinea a lady's, to be paid at the first lecture, when the hours of attendance for the future lectures will be fixed agreeable to the majority of the subscribers" (Hans 1951, 146–48).

Mr. Adam Walker, a teacher of the belles lettres from Manchester, had traveled throughout northern England, Scotland, and Ireland since 1766. That was the year that he had purchased the scientific apparatus of William Griffith, a famous itinerant lecturer. Like his predecessor, Walker moved from town to town teaching two-week science courses. Typically they consisted of twelve two-hour lectures on topics such as astronomy, optics, magnetism, electricity, mechanics, hydrostatics, pneumatics, chemistry, and the general principles of matter. He made a decent living from ticket sales.

Two days later the course began. With his assistant nearby, Walker stood in front of a room packed with listeners from many walks of life. There were well-dressed barristers and barons interested in deepening their credentials as men of the Enlightenment, ladies determined to benefit from one of the only educational experiences available to them, businessmen who needed to keep abreast of sweeping technological changes, and simple workers and mechanics wishing to improve their skills.

Convinced of the seriousness of his mission, Walker sought to impress the same upon his audience. "Although [natural] philosophy has of late been branded as the cause of mischief by those whose interest is to promote ignorance and slavery in the world," he began, gesturing emphatically, "the wise and virtuous know well there is no inquiry that points out a more rational path to a knowledge of the Deity, or more rationally weans the mind from narrow and confining prejudices!" (Musson and Robinson 1969, 105).

Many nodded approval as Walker made his opening declaration. Eyes opened wider, and the intellectual adventure began.

THE SCIENTIFIC ORIGINS OF INDUSTRIALISM

Adam Walker was part of an educational movement that was sweeping through England during the last third of the eighteenth century. Lectures about complex technical subjects had been growing in popularity since the early 1700s when a first generation of lay educators began to spread the word. While the church and its Aristotelian supporters on the continent waged a rearguard struggle against unorthodox trends in natural philosophy, English men of the cloth welcomed a new learning. God's existence was not challenged by calculus and physics; rather, it was confirmed by them. As Francis Bacon put it: the work of God complements the word of God. The laws of nature were not the object of intense academic controversy or political persecution; rather, they represented an established consensus that was studied openly by all respectable people. Science, in short, was the rage in England.

This significant intellectual development had its origins in the 1680s at Cambridge University. It was there that a brilliant young academician, Sir Isaac Newton, synthesized the terrestrial physics of Galileo and the celestial physics of Kepler to found a new mathematically demonstrated physics based on the universal law of gravitation. Gravity was the attractive centripetal force that pulled the planets toward the sun in the same manner as falling bodies are pulled down on earth. It was balanced by the force of inertia, that is, the tendency of an object like earth to continue moving in its previous line of motion. The result was an elliptical orbit around the sun determined by the forces of gravity and inertia. Newton determined, furthermore, that gravity's force was always equal to a constant times the product of the masses of the two attracting objects divided by the square of the distance between them. Perhaps most significant of all was Newton's belief that these forces of motion in the universe were the work of an omniscient and omnipotent God. In the Newtonian synthesis, there was no conflict between science and religion (Jacob 1988, 60).

Although it is always difficult to identify—let alone isolate—cultural factors that influence scientific motivation, historians agree that Newton and his kindred spirits at Cambridge were probably responding to a constellation of troubling developments in the world of seventeenth-century science. First and foremost was the work of René Descartes. Equipped with a powerful intellect, the French mathematician shred Aristotelian logic to ribbons with his mechanistic assertions that there were no "forms" or special "qualities" in matter; rather, everything in nature, from the human body to the stars, was simply matter in motion. "The rules of mechanics ... are the same as the rules of nature." Such truths were visible to all, he claimed, who proceeded from one reasoned deduction to another without succumbing to the old shibboleths. Descartes himself believed in God—what alarmed Cambridge circles were the uses of Cartesian science in the hands of others. Continental believers in magical, occult forces in nature, such as the followers of Benedict de Spinoza, seized upon Descartes' challenge to established religion and hence political authority to win over the lower classes for republican and revolutionary agendas. The frightening potential of these views was demonstrated in England during the chaos of the civil war (1642–1649) and turmoil of the republican regime of the 1650s. Others used Descartes to justify absolute monarchy, for if material and mechanical factors operated in nature to the exclusion of spiritual forces, then only a powerful head of state could prevent the victory of greed, cynicism, and self-interest. The absolutist pretensions of Charles I and resulting bloodshed of the 1640s, followed by the potential for a repetition of these troubles under his pro-Catholic grandson James II in the 1680s, further discredited Descartes in the eyes of moderate Protestant scientists in England. By reinserting God into the laws of the universe, Newton and his compatriots sought to promote religiosity and parliamentary order at the expense of paganism, materialistic atheism, and the whimsical, irresponsible, "popish" rule of kings.

The ouster of James II during the "Glorious Revolution" of 1688 freed parliament from absolutist challenges and the Church of England from Catholic intrigues. It was thus quite logical and consistent that Newtonian physics began to sound from the pulpits of the most distinguished congregations in London. Anglican clergymen like Richard Bentley, Samuel Clarke, and William Whiston preached that "the same God whose laws of motion Newton had discerned in the natural world would also inevitably ensure order, prosperity, and the conquest and maintenance of empire in the political world," writes Margaret Jacob (1988, 96). Interestingly enough, Whiston was among the first to take this message into more secular settings. In the 1690s and early 1700s he joined other popularizers of the new physics in London by holding lectures in coffeehouses, printers' shops, and private homes. Like the sermons, these talks avoided the high math that laymen did not understand. Unlike the pulpit lectures, however, Whiston's private tutorials were more technical, involving modular demonstrations of Newtonian physics in fields like mechanics and pneumatics. The first private lectures were therefore of practical interest to the merchants and craftsmen who attended. And so the great man's ideas started to permeate London society "top down" with the sanction of established authority (Stewart 1992).

Undoubtedly the best-known popularizer of Newtonianism in the early 1700s was Jean Theophile Desaguliers, a French Huguenot refugee who served as official experimenter of the Royal Society of London. Desaguliers began to lecture at his home in Westminster in 1713, and later at the Bedford Coffee House in Covent Garden. He was also one of the first to venture into the provinces, teaching at the famous Gentlemen's Society of Spalding, the earliest scientific society founded outside of the capital. Desaguliers believed that scientists had to rise above mere contemplation of the works of God in nature "to make [artistry] and nature subservient to the necessities of life" (Musson and Robinson 1969, 40). Entrepreneurs, in turn, needed to be familiar with the details of mechanics' jobs as well as the general and specific scientific principles behind the working of engines and machines. In one typical course, Desaguliers moved from a theoretical discussion of the elements, matter, gravity, inertia, and motion, to machines: "The whole effect of mechanical engines, to sustain great weights with a small power, is produced by diminishing the velocity of the weight to be raised, and increasing that of the power in a reciprocal proportion, of the two weights, and their velocities" (Jacob 1988, 143). He avoided obtuse mathematical explications, preferring to show with models how these principles applied to levers, weights, pulleys, wedges, waterwheels, and steam engines. For his decades of service and illumination he received the Royal Society's Copley gold medal in 1741.

Through Desaguliers we can trace connections to other important avenues along which the new mechanical knowledge spread. The Spalding Gentlemen's Society, founded in 1712, was soon emulated in Manchester, Birmingham, Northhampton, Newcastle, Edinburgh, and many other towns. All brought men of wealth and learning together to learn about science and investigate its application to farming, canal building, and industrial production. The most successful societies welcomed expert lecturers, received the published transactions of the London Royal Society, amassed well-stocked scientific libraries, and even conducted their own experiments (Morton and Wess 1993; Stewart 1992; Jacob 1988; Musson and Robinson 1969).

Desaguliers was also a founding member of English Freemasonry. He joined the Horn Tavern Lodge in Westminister, one of four that merged in 1717 to form the Grand Lodge of London. British Freemasons were closely linked to the Newtonian ideology that was coming to dominate the English Enlightenment. About a quarter of the London brothers belonged to the Royal Society, while the bulk of the membership was drawn from the same mercantile class that sat in parliament, attended Anglican church services, and imbibed the scientific sociability of the coffeehouses. "They seek to determine the causes of things with grace and without arrogance, affectation, or prejudice," as one early brochure described the Masonic brothers. "One is accustomed there only to science, [and] a thirst for knowledge. . . . Truth is their ultimate goal and reason is the leader they follow." Masons also learned, the booklet continued, "how one should use the things which are created" (Keller 1918, 13). Accordingly, some lodges accumulated impressive scientific-book collections and featured lectures and courses on physics or mathematics. Like the scientific societies, Freemasonry spread from London into the midlands and provinces during the 1720s and 1730s (Hall 1937; Jacob 1981, 1988).

The circulation of scientific information in England accelerated during the middle third of the eighteenth century. This was largely the work of a growing number of pioneering educators. One of the first was John Horsley, a Presbyterian minister from Morpeth, near Newcastle, who opened a private school around 1730, which gave students an opportunity to study experimental natural philosophy utilizing the latest models and apparatus. Horsley also gave special lectures for mine owners in the Newcastle region. "All possible pains will be taken in these lectures," noted one advertisement, "to render everything plain and intelligible even to those who have no previous acquaintance with mathematical learning" (Stewart 1992, 367). The same approach was taken by Caleb Rotheram in the 1740s at Kendal Academy in Manchester. Science, which he dubbed "one of the most useful and entertaining branches of learning ... and of great importance in the common affairs of human life" (Musson and Robinson 1969, 102), was a regular part of the curriculum there. The example was followed in London, Birmingham, Sheffield, Leeds, and other towns.

Reinforcing these educational tendencies were the itinerant lecturers who began to teach special courses in science. The first to devote himself to this task was probably James Ferguson, a self-educated Scotsman who took to the road around London in the late 1740s. William Griffith was another very popular lecturer in the southwest in the 1750s, and he was followed in the 1760s by Adam Walker. And there were many more: John Arden, Henry Moyes, John Warltire, Henry Clarke, Benjamin Donne, John Banks, Peter Clare, and others. Focusing on the single town of Manchester, Professors Musson and Robinson found that "almost every year for the last forty years of the eighteenth century Manchester was visited by one and sometimes by several of these lecturers, each usually giving a course, occasionally two courses, of lectures numbering sometimes as many as thirty, spread over several weeks or even months" (ibid., 102–103). When Walker visited York in 1772, he claimed—evidently in good faith—that six hundred students had recently taken his course in Manchester and Liverpool. His wagon with its strange-looking contents would become more familiar-looking to the inhabitants of the north country as the popular educational movement widened and deepened.

By the eve of the Industrial Revolution—indeed, well before it—England could boast of a significant transformation in the way its countrymen thought about mechanical problems. There was a new mode of thinking, a widespread acceptance of machines, and an impressive understanding of the technical principles that lay behind the working of mechanical devices. Looking back to the beginning of the century, one contemporary concluded that it was the "general diffusion of scientific knowledge among the practical mechanics of this country, which has, in a great measure, banished those antiquated prejudices, and erroneous maxims of construction, that perpetually mislead the unlettered [artisan]" (ibid., 103). Even the ordinary millwright, observes historian David Landes, was usually "a fair arithmetician, knew something of geometry, leveling, and [measurement], and . . . could calculate the velocities, strength, and power of machines: He could draw in plan and section" (1969, 63). Perhaps even more important was the sophisticated grasp of eighteenth-century science by English merchants and producers large and small who had to make tough investment decisions. The rise of England as a Great Power brought conquests and new market demands as the century unfolded. These developments underscored the inadequacy of the old technological system. Moving beyond known technologies, however, had to be more than a leap of faith, for feelings and emotions usually strengthened a conservative attachment to the old ways. Rather, one needed a science that was powerful enough to predict outcomes and illuminate the difference between various technological options. And England had this when the country needed it.

THE ECONOMIC ORIGINS OF INDUSTRIALISM

After a harrowing time of crop failure and Indian wars in the seventeenth century, England's North American colonies began to grow and prosper. Population rose from about 50,000 in 1650 to 251,000 in 1700, and then quadrupled to nearly 1.2 million in 1750. The number of English colonists would double again to 2.1 million by 1770. By this latter date there was one American colonist for every three Englishmen in the home country. The three largest American cities grew from towns of a few thousand in 1650 to forty thousand (Philadelphia), twenty-five thousand (New York), and sixteen thousand (Boston) in 1775. With the so-called Navigation Acts limiting manufacturing and shipping in the colonies, English businessmen were the major beneficiaries of a rapidly expanding market. Thus the average annual value of exports of British manufactured goods to the thirteen colonies skyrocketed from 270,000 pounds sterling in the early 1700s (1701–1705) to 2.2 million in the early 1770s (1768–1774). This was almost a sixth (15.4 percent) of total British exports (Bruchey 1968; Jensen 1950; Mathias 1969).

Political developments in England, meanwhile, had a major impact on the expansion of the British Empire—and hence on sales of British manufactured wares in other parts of the world. The expulsion of James II in 1688 brought greater decision-making powers for the House of Commons. The new king, William III of the House of Orange, reigned as a "parliamentary monarch" who abandoned the efforts of his predecessors, the Stuarts, to rule "absolutely" without the cooperation of elected representatives. As a Dutchman, moreover, William of Orange was devoted to Protestantism, not Catholicism; attuned to the needs of the merchant class, not landed aristocrats; and committed to checking French expansionist designs in Europe and the world, not abetting them. The result was a unique political cooperation between the king and the wealthy merchant and banking elite that controlled parliament. Previously unwilling to lend their expanding riches to monarchs who often refused to repay—and were perceived, in addition, as poor guardians of the national interest—the English bourgeoisie now endorsed and facilitated the government's credit needs (Dickson 1967).

The practice of government borrowing was institutionalized in 1694 with the establishment of the Bank of England. It was founded during the War of the League of Augsburg (1689–1697), William's continuation of his struggle against the French. The bank's first charge was to lend the state 1.2 million pounds. Its vital intermediary role of funneling the nation's growing wealth into government coffers grew during a second round of fighting, the War of the Spanish Succession (1702–1713). By war's end the national debt of England stood at an impressive 37.4 million pounds. For the first time in modern history a nation had been able to spend vast sums of money far in excess of its tax revenues. After the War of the Austrian Succession (1740–1748), the state owed its loyal business supporters 78 million pounds—a debt that doubled to over 150 million pounds after the worldwide Seven Years' War (1756–1763). During this climactic showdown with the financially strapped absolute monarchy of France, Great Britain was able to afford fleets and armies in four theaters of action plus pay huge subsidies to its allies. Three-quarters of a century after the Glorious Revolution, England had added Canada and Florida to its North American possessions; reduced the French to a secondary power in the Caribbean, the Mediterranean, and India; and thoroughly frustrated two great French kings in their quest to dominate continental Europe. As P. G. M. Dickson (1967) reminded us, Britannia ruled the waves—on credit (Tracy 1990, 1991; O'Brien 1991, 1993, 1994).

The British mercantile class reaped gargantuan returns on its investment in the state. According to recent estimates the value of British exports rose from 3 million pounds in 1700 to 28.4 million in 1800. Next to the dynamic North American market, the largest growth areas were the Caribbean, Africa, and the East Indies. Altogether these extra-European markets expanded over fivefold during the 1700s, rising rapidly from 14.7 percent of total exports to a remarkable 69.4 percent. These gains were reflected most graphically in the rapidly expanding output for export of iron, woolen, cotton, and other textile goods. Using an index number base of 100 for the year 1700, export industries grew to 544 during the eighteenth century. Probably more than half of British textile production was exported. Although iron was made largely for domestic markets, some towns like Birmingham and Sheffield produced mostly for foreign customers. When viewed in terms of the percentage of overall industrial output sent abroad, on the other hand, the trend seems somewhat less dramatic—a fact that one school of skeptical economists seized upon (see below). The percentages produced by "new economic historian" N. F. R. Crafts, for example, rose from 24.4 in 1700 to 35.2 in 1760, but then slipped to 21.8 in 1780 before rising again to 34.4 in 1801. When merchandise exports are viewed as a ratio of national income, moreover, the trend also appears more gradual: over an entire century the figure ascended from 5 to 6 percent to only 14 percent (Deane 1967; Deane and Cole 1969; Mathias 1969; Crafts 1985; Engerman 1994).

Using the same index numbers, manufacturing output oriented to the domestic market rose more slowly from 100 to 152 in the eighteenth century. We should not underestimate the significance of this 52 percent increase in home industries, however, for in absolute terms, sales of industrial commodities in England were far greater than in the market abroad, absorbing the overwhelming majority of manufactured goods produced (Hudson 1992; Mathias 1969). This expansion of output resulted from a number of important developments. After remaining constant in the late 1600s, and then beginning to rise in the early 1700s, population increased almost 40 percent between 1740 and 1790. It is significant, however, that national income rose about 150 percent after 1700, stimulated somewhat by agricultural improvements (see below), but more so by commercial expansion and industrial activity. To an informed observer like Adam Smith, the Scottish economist whose famous *Wealth of Nations* appeared in 1776, workers' average real incomes seemed to be growing faster than the population by the 1770s. "The real quantities of the necessities and conveniences of life which are given to the laborer have increased considerably during the present century" (Deane 1967, 88). Contemporary scholarship points much more to a jump in fashion-driven consumption "among the middle class and the class of small tradesmen," while finding "little evidence of rising real incomes of the mass of the population until the early nineteenth century, and even less of participation in revolutionary fashion-orientated consumption" (Berg 1994, 129, 128). Far from rising, in fact, wages declined in regions like the West Country that were left behind by expansion in other areas, creating impoverishment and considerable resistance to new technology. Other historians, while conceding that there was no trend in rising lower-class wages to enhance domestic demand, assert nonetheless that the "industriousness" of women and children probably salvaged family incomes (De Vries 1994). Distaff spinning, hand knitting, lace making, and straw plaiting provided monies that allowed purchases of household textiles, clothes, cutlery, pottery, and basic furniture (Hudson 1996). Road-building schemes, canal construction, and other transportation improvements completed the process of creating a national market in eighteenth-century England. As we have seen from the statistics, producers responded to these new opportunities by expanding output.

But it was not an easy proposition. Those English merchant-manufacturers most attuned to money-making opportunities—and therefore most sensitive to profits unmade because of productive blockages—responded initially to the market challenge by widening the so-called domestic system of manufacturing. Under this organization of labor, merchants familiar with the market purchased raw materials like wool, flax, and cotton for distribution in the countryside to small-town craftsmen and country dwellers in need of extra income. By the early 1700s, Lancashire and the West Riding area of Yorkshire were famous for this cottage industry, as were East Anglia, Kent, and the southwestern counties stretching from Devon, Somerset, and Dorset into Hampshire. As the century unfolded, these areas of domestic industry prospered: output of woolen cloth in the West Riding area, for instance, grew from 100,000 pieces in 1740 to 180,000 in 1780. But clothiers' complaints also multiplied, for it was increasingly complicated to coordinate the distribution of large amounts of material; rising transportation costs cut into profits; and theft of raw materials was nearly impossible to detect. It made good business sense, therefore, for some entrepreneurs to consider concentrating all factors of production under one roof in "factories."

At the same time, enterprising cottage craftsmen began to adopt some of the first ingenious spinning devices (see below). Such "proto-industrial," risk-taking innovation, in turn, prompted the merchant factory owners to follow suit, mechanize, and increase productivity to meet demand. Recent research has demonstrated, furthermore, that other kinds of producers were responding to the same market stimuli. Local small-town or urban craftsmen, operating sometimes in a single-family operation, at other times in artisan cooperatives, or even in rudimentary factorylike settings featuring specialization and a division of labor, also attempted to penetrate the new markets. Historians now believe that a great deal of England's productivity gains occurred in these "flexible specialization" and rural proto-industrial settings. "Small producers offered economic advantages of creativity, nimbleness and easy entry; they could develop 'flexible technologies', easily changed across products and activities, as well as skill-intensive processes and a range of product choices for localized and regional tastes," writes Maxine Berg (1969, 63). Emphasis on this "polymorphic nature" of the early Industrial Revolution—the coexistence and co-mingling of the first factories with small-scale artisan production as well as cottage industriousness—is the emerging historical consensus (Landes 1969; Mantoux 1961).

England's economic crisis was multilayered, however. At the same time that manufacturers and other producers attempted to accelerate output, either by extending the putting-out network or investing in the first factories, a new "supply side" dilemma emerged—the movement to "enclose" farms. Eager to apply the new scientific principles to agriculture, landowning gentry petitioned parliament to redistribute the divided strips of land (which characterized medieval agriculture) into consolidated—or enclosed— farms. Each county required a separate act of parliament. There were only nine such acts of enclosure between 1702 and 1730, about three hundred thirty between 1730 and 1780, and nearly seventeen hundred between 1780 and 1810. Because enclosure required many workers to build fences and barns, labor became as scarce as trees and charcoal. Decades ago J. D. Chambers (1953) and, later, François Crouzet (1967) attached particular economic significance to this labor shortage. Crouzet writes:

> The innovations in agriculture, far from creating unemployment, were stimulating demand for labor, which contradicts the older view that England was at this period suffering from large-scale rural underemployment and that there was a mass exodus from the countryside to the industrial centers. We must also remember that up to mid-century the population grew only very slowly. Although it then started growing fast, a number of years had to pass before a large supply of hands was available for the labor

market.... There was therefore a relative shortage of labor in industrial districts, as is proved by the quite sharp rise in money wages there... during the first half of the eighteenth century (ibid., 170).

Higher wages in agriculture and industry fueled purchasing power, Crouzet argued, which added even more "demand-side" pressure on entrepreneurs to increase production. In recent years historians have downplayed his purchasing-power thesis, concluding that a moderate 6 percent of increased expenditure on manufactured goods could be attributed to agricultural productivity-induced higher wages in town and country (O'Brien 1985). Moreover, the new research highlights the temporary nature of enclosure's demand for increased labor, pointing to farms, for instance, that used only a third as much labor *after enclosure* as under the open field strip-farming system. In the long run, in other words, enclosure was "labor releasing," not "labor using," although it seems clear now that the labor released in the long run tended more often than not to stay in the countryside, contributing thereby to rural poverty (Berg 1994, 88–89). Nevertheless, the arguments of Chambers and Crouzet still illuminate the particular circumstances that triggered industrial growth, especially after 1780—the very same decades of the accelerating enclosure movement. Faced with labor shortages and growing demand in the late 1700s, producers intensified their search for better manufacturing technology.

Indeed the rush to increase production for both foreign and domestic markets placed even more pressure on the technological system. Technology had run up against severe constraints in Europe during the 1600s. England experienced these limits earlier and more intensely as trees grew scarcer, charcoal costlier, and coal mines deeper and more difficult to drain as cost-effectively as in Europe. Consequently, the iron and steel required for more durable and effective machines were very expensive. Now, just as these older constraints were tightening, England faced challenging new demand-side problems created by rapidly increasing market possibilities at home and abroad and supply-side difficulties resulting from enclosure and a putting-out system that was approaching the limit. As industrial production strained to increase, the technological system threatened to break down.

It was thus very fortunate for England that scientific ideas had begun to spread from London to the provinces during the first years of the century. There are never any guarantees in the history of technology that a society in need of new productive techniques will be able to create them. As noted above, England's scientific, intellectual, and educational revolution provided the nation with the necessary reservoir of skills as it attempted to improve the old technological system. After describing the evolution of this new way of producing in the next passage, it will be necessary to consider some recent challenges to traditional ways of viewing the industrial revolution and more recent responses to these revisionist positions.

ENGLAND'S TECHNOLOGICAL REVOLUTION

In one of the more memorable passages about the history of technology, David S. Landes writes:

> In the eighteenth century, a series of inventions transformed the manufacture of cotton in England and gave rise to a new mode of production—the factory system. During these years, other branches of industry effected comparable advances, and all these together, mutually reinforcing one another, made possible further gains on an ever-widening front. The abundance and variety of these innovations almost defy compilation, but they may be subsumed under three principles: the substitution of machines—rapid, regular, precise, tireless—for human skill and effort; the substitution of inanimate for animate sources of power, in particular, the introduction of engines for converting heat into work, thereby opening to man a new and almost unlimited supply of energy; [and] the use of new and far more abundant raw materials, in particular, the substitution of mineral for vegetable or animal substances. These improvements constitute the Industrial Revolution [of the eighteenth century] (1969, 41).

Notice that Landes was very familiar with the "systemic" nature of these technological changes. Many industries "effected comparable advances" that were "mutually reinforcing" and that, in turn, "made possible further gains on an ever-widening front." As we know from chapter 1, Bertrand Gille articulated a model in the late 1970s of technical structures, ensembles, and concatenations that, taken together, functioned as a technological system to meet society's needs. In the section below, I first identify the eighteenth-century technological breakthroughs that, according to Landes and Gille, enabled England, faced with those economic challenges described above, to develop the initial component parts of a new system. Next, I consider some of the reinterpretations and challenges to the standard model of the Industrial Revolution as well as more recent attempts to revitalize this standard model. After this historiographical digression we revisit continental Europe to investigate the reasons why nothing comparable occurred there in the eighteenth and early nineteenth centuries. Finally, we will trace the evolution of the mature, fully developed technological system that made England the "workshop of the world" by the 1830s and 1840s.

* * *

The first mechanical innovations in the textile industry of this period were designed for wool, and with good reason. Consumption of raw wool had increased 50 percent between 1700 and 1740 and then rose another 33 percent by 1780. The value of woolen production in 1783 was 77.2 percent of the entire English textile industry, towering over linen (18.4 percent) and cotton (4.4 percent). John Wyatt and Lewis Paul responded to the challenge by patenting their spinning frame (1738), an ingenious combination of rollers and flyer mechanisms that spun fibers onto twenty-four spindles. John Kay also intended his fly-shuttle (1733) as a machine to weave wool. By means of a cord, the weaver could "shoot" the loom's shuttle through the warp and back again, thus enabling one man to perform work formerly requiring two men—and considerably speeding up the weaving process (see figure 2.1).

Science and Technology in the World's Workshop 43

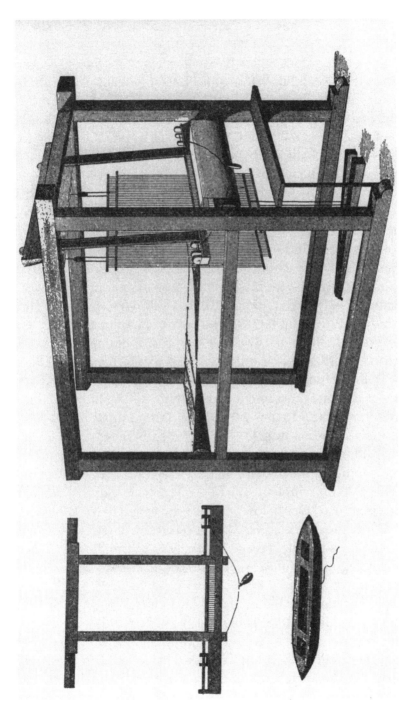

Figure 2.1: Kay's fly shuttle showing complete loom (*right*) and batten, operating cord, and shuttle (*left*). From Charles Singer, ed., vol. 3 of *A History of Technology* (1958), p. 170.

It is interesting, however, that both mechanisms came slowly into use in the 1740s and 1750s for cotton rather than wool. The raw material that dominated British textiles was relatively weak, uneven in texture, and therefore difficult to spin and weave with the awkward, jerky machines of the early Industrial Revolution. Cotton, in contrast, was stronger and more consistently resistant to the abuse of machines. It was also the rage fabric, popular among all classes as a cheap, light, washable garment. This is most persuasively demonstrated by the volume of net cotton imports, which took off from around a million in 1700 to 3.7 million in 1770. By 1800 they had catapulted to 47.2 million (Landes 1969; Mathias 1969; Paulinyi 1991; Usher 1967).

Cotton's meteoric rise in popularity made it imperative to find new methods of production. Indeed it is no exaggeration to describe the changes in cotton manufacture that took place roughly between 1765 and 1790 as revolutionary. First came Richard Arkwright's water frame (1769) and James Hargreaves's spinning jenny (1770). Both were variations of Wyatt and Paul's spinning frame, but whereas the earlier device produced perhaps twenty-five times as much as the spinning wheel in one hour, the jenny (in its later designs) was eighty times and the water frame several hundred times as productive (see figure 2.2). They were joined by Samuel Crompton's spinning mule (1779), so called because it was a technical hybrid of the jenny and the water frame. Doing two hundred to three hundred times the work of the old spinning wheel, the mule soon relegated the simpler jenny to use in cottage industry, where it had first been used (see below), then by the early 1800s, to obsolescence. These machines placed strain on weaving, which had improved little since the spread of Kay's fly shuttle in the mid-1700s. The technological response was Edmund Cartwright's power loom, the recipient of successive patents between 1785 and 1788. There were about twenty-four hundred of these looms in England by 1813 (Landes 1969; Mann 1958; Paulinyi 1991; Usher 1967).

The improvements to cotton spinning and weaving described above were certainly the most famous that occurred in the textile industry. But the merchants and manufacturers who invested in

Science and Technology in the World's Workshop 45

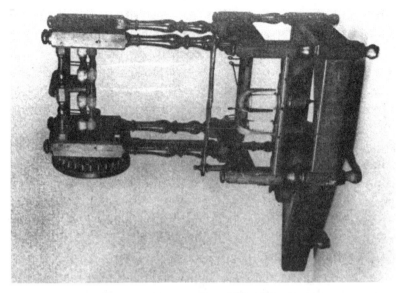

Figure 2.2: Spinning jenny (*left*) and water frame (*right*). Courtesy of Science Museum, London.

these machines desired change in every stage of production in the textile "concatenation" (see below). Many attempts were made after mid-century, for example, to mechanize the carding, drawing, and roving—that is, preparatory—processes of raw cotton. But it was Cartwright who succeeded with a series of patents between 1775 and 1785. Easily the most dramatic breakthrough in the finishing stage came in bleaching with the increased use of sulfuric acid. Traditionally, bleachers used sour milk or simply spread cloth in sunlit fields. But as Landes writes, "it was no longer feasible to bleach cloth in open meadows when more of it was being turned out than there was ground available" (1969, 87). The solution to this "technological bottleneck" was found in 1746 by John Roebuck, owner of a metal refinery in Birmingham. Chemists made sulfuric acid expensively (and in small quantities) by burning sulfur and potassium nitrate to produce gases that were then absorbed into water held in glass containers that resisted the corrosion of the acid. Having a scientific background, Roebuck knew that sulfuric acid did not act on lead. By burning sulfur and saltpeter inside large lead-lined vats containing a shallow depth of water, his "lead chamber" process succeeded in manufacturing greater quantities of sulfuric acid at vastly reduced prices. The technique spread quickly during the 1750s and 1760s (Clow and Clow 1958; Mann 1958; Paulinyi 1991).

The second overall category of technical change during the English Industrial Revolution was the application of steam power. We have already observed in chapter 1 that Thomas Newcomen produced an "atmospheric" steam engine for use in the mines of Wolverhampton in 1712 (see figure 2.3). It operated by introducing steam into a cylinder, then cooling it to create a vacuum that sucked the piston downward. The piston, in turn, lowered one end of a beam, while the other end, drawn upward, activated pumping mechanisms below. Gravity drew the raised end of the beam down again, thus returning the cylinder and piston on the other side to their starting positions. The Newcomen "fire engine" was weak, thermally inefficient, slow, cumbersome, and expensive. But, with fifty to sixty horsepower, it had three to five times the power of early waterwheels, pumped four times as much water in a day as teams of

Figure 2.3: A Newcomen steam engine, 1717.
Courtesy of Science Museum, London.

horses, and did so at 80 to 85 percent of the cost. There were well over four hundred Newcomen steam engines in England by 1775, mostly concentrated in the coalfields of Cornwall and Northumberland (Ferguson 1967b; Pacey 1980; Troitzsch 1991).

By this time two of England's best engineering minds were working on improvements to atmospheric steam engines. The first was John Smeaton, an instrument maker from Leeds. Smeaton's systematic experimentation with larger-scale Newcomen engines in the late 1760s resulted in a doubling of their thermal efficiency. The second was James Watt, an instrument maker from Glasgow. Watt's contribution from 1765 to 1785 was more important, for he actually transformed the steam engine into a superior mechanism (see figure 2.4). First, he added a condenser that was separate from the cylinder. By allowing the cylinder to remain hot, Watt doubled the efficiency of the best Smeaton engines. He also incorporated fly-ball governors that regulated engine speed. As the engine accelerated, the balls moved outward with centrifugal force. Through a series of connecting rods, this same action released steam from a valve, thus reducing pressure and engine speed. Finally, Watt introduced pistons that moved up and down in reciprocal motion. He accomplished this by injecting steam alternately into both ends of the cylinder. A connecting rod and crank converted the upward and downward movement of the beam into the rotary motion of a wheel. It was now practical to run machinery with steam engines. There were over five hundred engines of this design in England by 1800, most of them driving mule spinners, power looms, mine pumps, and ironworks. There were around twenty-two hundred steam engines of various designs in England at this time (Ferguson 1967b; Paulinyi 1991).

The final technological breakthroughs in Landes's schema resulted from the substitution of mineral for vegetable sources of energy. Hard-pressed to find fuel cheaper than wood and charcoal, English manufacturers had relied increasingly on coal for many centuries. While this substitution proved feasible in brewing, glass making, nonferrous metallurgy, and other branches of production, it remained impractical for ferrous metals. In iron making the fuel

Science and Technology in the World's Workshop 49

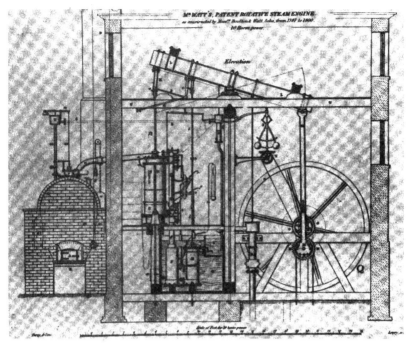

Figure 2.4: Diagram (*top*) and photograph (*bottom*) of the Watt engine. Notice the connecting rod for conversion of alternating to rotary motion. Courtesy of Science Museum, London.

came into direct contact with the ore, allowing impure elements common to most coals ample opportunity to combine with iron. If present in even minute quantities, for instance, sulfur, phosphorous, manganese, and silicon created either a "red short" iron that crumbled during hammering or a "cold short" finished iron that was weak, brittle, and porous. For these reasons, charcoal, a vegetable fuel free of impurities, remained the ironmaster's preference.

Abraham Darby, an iron maker from Coalbrookdale in Shropshire, began to solve this dilemma as early as 1709. Company records indicate that he succeeded that year in smelting iron ore with coke, that is, coal that had been roasted in ovens to burn away impurities. Forty years later, however, Darby remained the only iron maker in England using the process. Despite the fact that he coked Shropshire coals, which were unusually low in sulfur, quality problems persisted. The cost advantage of using cheaper mineral fuel was not significant enough, moreover, to offset the greater difficulty of working with a qualitatively inferior iron. Only the steep increase in charcoal prices and the growing demand for iron after 1750 made coke pig iron a more attractive alternative. By 1785 England produced 14,000 tons of charcoal pig iron and 47,700 tons of the coked variety (see figure 2.5) (Brose 1985; Hyde 1977; Tylecote 1976).

The use of coal in the conversion of high-carbon pig iron to malleable, low-carbon wrought iron occurred simultaneously in the decades after mid-century. Patented between 1761 and 1773, the so-called potting-and-stamping process produced a useable finished iron. The crude-iron pigs were reheated to burn off some of the impurities, then broken into small pieces and placed into clay pots with a limestone "flux." The crucibles protected the iron from the deleterious coals and gases of the fire as high temperatures oxidized the carbon and facilitated the absorption of sulfur and other impure elements into the flux. By 1788 England turned out 16,400 tons of charcoal wrought iron and 15,600 tons of coke-bar iron.

Ultimately more important was the so-called puddling-and-rolling process patented by Henry Cort in 1783–84—but not perfected until the 1790s. Puddling utilized reverbatory or indirect heating. The coals were separated from the pigs by a low wall, which

Science and Technology in the World's Workshop 51

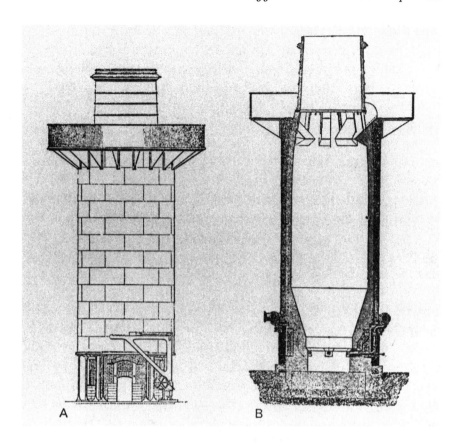

Figure 2.5: Scottish blast furnace of the early 1800s (*left*) with cross section (*right*).
From Charles Singer, ed., vol. 4 of *A History of Technology* (1958), p. 112.

allowed heat to rise over the divider and convert the iron (see figure 2.6). A worker stirred the iron mass with a metal bar to accelerate the oxidation of carbon. The iron then passed through a rolling mill, which replaced the tedious hammering stage. English bar-iron output surpassed a hundred thousand tons in 1805 as puddling gradually supplanted both charcoal and potted and stamped wrought iron. Coke pig iron production soared over eighty thousand tons in the early 1790s, reaching two hundred fifty thousand tons by 1805 (Hyde 1977).

It is important to emphasize that the trend toward mechanization, steam power, and coke iron was just beginning. Thus the mechanical innovations in cotton textiles had not spread to any appreciable degree before 1815 into the woolen industry. Rather, this sector expanded along the old lines of the putting-out system. In West Riding, for example, domestic production of woolen cloth rose from 180,000 pieces in 1780 to 430,000 pieces in 1803. The technological revolution in cotton production also remained incomplete. Power looms were still awkward and clumsy, even for a tough material like cotton. Fiber breakage was so common, in fact, that power looms were hardly more productive than the hand-powered fly shuttle. As significant as steam engines were for the history of industrialization, furthermore, they remained an auxiliary source of power in the late 1700s. There were easily five to ten waterwheels for every steam engine in England at this time, and water power generated over twice as much total horsepower. Indeed research and experimentation by men like Smeaton improved the efficiency of waterwheels until "overshot" designs (where water hit the top of the wheel) generated two hundred horsepower. The age of coke iron, too, was just starting. Modern tests on Coalbrookdale iron from the late 1700s—the best coke iron in England—revealed quantities of sulfur, phosphorous, manganese, and silicon far too high for general-purpose iron that had to be malleable enough for hammering without crumbling, strong enough to withstand stress and strain, and texturally pure enough to endure the effects of repeated heating and cooling. For these uses England continued to produce charcoal iron and import tens of thousands of tons from

Science and Technology in the World's Workshop   53

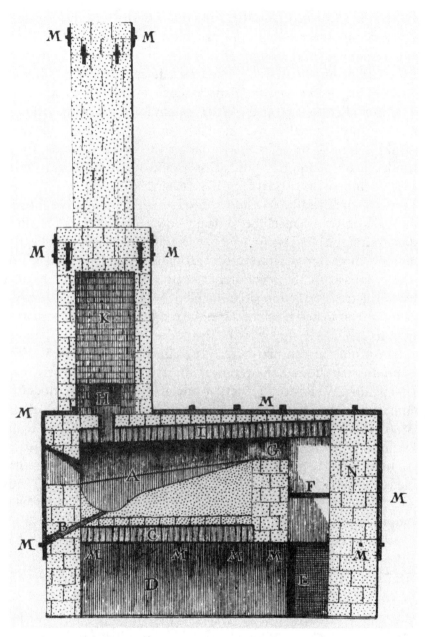

Figure 2.6: Puddling furnace showing compartment for coals (F), dividing wall (G), compartment for reducing iron (A), and puddler's rod (B). From Gabriel Jars *Voyages métallurgiques* (1774), p. 75.

Sweden and Russia. Coke iron tended to be used for thick castings like vats, pipes, cannon, and bridge and building components that either required little strength or hardness or where the thickness of the casting compensated for coke iron's defects. In short, although England began to devise the parts of a new technological system, it was compelled to exploit the older system for most of the nation's needs. Only gradually over the half century from 1800 to 1850 would it be able to alter this situation (Brose 1985; Hyde 1977; Landes 1969; Mantoux 1961; Musson and Robinson 1969; Pacey 1980; Paulinyi 1991; Tylecote 1976; Usher 1967).

Such qualifications aside, however, the remarkable breakthroughs in textile machinery, steam power, and ferrous metallurgical technology had begun to shape England into the first industrial nation. An earlier generation of historians certainly viewed it this way, emphasizing the revolutionary nature of the new era. A great technical transformation accompanied by a spectacular "wave of gadgets" eliminated bottlenecks in the productive process, increased output, and accelerated growth. The author of the "gadget" phrase, T. S. Ashton, claimed that "after 1782 almost every statistical series of production shows a sharp upward turn" (1955, 125). A few years later Walter W. Rostow (1963) formulated his famous model of "industrial take-off" using England's "sharp upward turn" of the 1780s and 1790s as the classic case (see chapter 3). Next Phyllis Deane and W. A. Cole (1969) published a pioneering statistical study of British economic growth that showed industry and commerce jumping from 0.49 percent annual increases during the 1760s and 1770s to 3.43 percent yearly growth in the 1780s and 1790s. Indeed England seemed to have "taken off."

RECENT REINTERPRETATIONS AND DEBATES: WAS THERE AN INDUSTRIAL REVOLUTION?

In recent decades economic historians and historians of technology have reconsidered the concept of an "industrial revolution" in England. Did this dynamic imperial nation really lurch dramatically into a period of rapid and accelerating industrial growth in the late 1700s? Did new technology function as the engine for impressive gains in productivity? Did the novel industrial sector spawn burgeoning towns and cities that quickly restructured the face of rural England?

In the 1980s economic historians began to answer no to these queries. Utilizing sophisticated new statistical techniques, N. F. R. Crafts and C. K. Harley produced new estimates that cast a skeptical light on the dynamic nature of England's industrial transformation (Crafts 1985; Harley 1982; Crafts and Harley 1992). Thus Crafts eventually settled on a startlingly low figure for annual industrial growth in the 1780s and 1790s—only 1.96 percent—while Harley scaled the numbers back even further to 1.5 percent for the entire period from 1770 to 1815. To reiterate, Deane and Cole (1969) had put late-eighteenth-century industrial growth at 3.43 percent. The new studies also deemphasized the economic impact of new technology, finding only slight shifts in the proportion of investment plowed back into industry, and only slight increases, therefore, in the productivity of capital and labor (i.e., total factor productivity)—a mere 0.1 percent growth per year from 1760 to 1801, and even more surprising, only 0.35 percent yearly from 1801 to 1831. So much, apparently, for "sharp upward turns" and convincing models of industrial "takeoff." Simultaneously, Crafts and other economic historians (Mokyr 1985) revised estimates of the growth of foreign markets during early British industrialization, concluding that foreign demand was a steady, not an increasingly important, factor in the seventeenth and eighteenth centuries—in other words, that the importance of exports in the late 1600s and early 1700s had been understated in earlier studies. They found, furthermore, that the ratio of exports to industrial output and national product actually declined from 1780 to 1840. Other historians joined the chorus, arguing that new eighteenth-century technology lifted restraints on production, thereby creating markets that heretofore did not exist. They emphasized supply-side factors, in other words, not demand-side factors (Davis 1979; Thomas and McCloskey 1981).

Other studies added to the overall reassessment. Thus E. A. Wrigley (1989) altered the historiographical landscape with his portrait of a traditional English economy that survived amid change well into the nineteenth century. He found the nation's population growth concentrated, for instance, in towns of small to medium size whose craftsmen and shop owners catered to the local market and rural hinterland. In the country, too, cottage industry predominated, not somewhat more sophisticated proto-industrial operations, let alone factories. Wrigley claimed that productive techniques remained largely stagnant in this relatively backward small-town/rural sector, one that employed 90 percent of all workers in town and country in the early 1800s. He argued, moreover, that England's traditional economy depended on "nature" for energy (wood and charcoal), power (human and animal muscle), raw materials (e.g., wool, flax, and dye crops), and, of course, food. The image of continuity and persistence of the old ways reinforced the historical impact of the lower estimates of Crafts and Harley and put all notion of an industrial "revolution" in perspective.

It did not take long, however, for the historiographical pendulum to swing back in the other direction. The most serious challenge to the new orthodoxy of Crafts, Harley, and Wrigley came with the publication of the second, substantially reworked edition of Maxine Berg's *The Age of Manufactures 1700–1820* (1994). This seminal work, already cited in the text above, shed a completely different light on Wrigley's traditional economy, with its allegedly backward techniques, by illuminating the technically innovative, market-oriented, and entrepreneurial aspects of production in small-town shops, artisan cooperatives, and proto-industrial cottage industry—the first to adopt the spinning jenny and the spinning mule, for example, were the small men.

> The merchants had no special price incentive to introduce the spinning jenny, though they did stand to gain from higher and readier supplies of yarn. The ones who stood to gain most were those who owned and ran the spinning jennies themselves. It was the cottage producers and those who ran small centralized workshops who reaped the first gains in efficiency from the jenny, and they did so until the merchants and factors saw the gains to be had through setting up their own larger jenny factories. It was not the

putting-out system itself which brought the jenny to the cotton manufacture, but a dispersed industrial structure which created opportunities for the direct producers to reap some of the gains from increasing their efficiency. The mule was only a development on the jenny, and it too was initially used at home and in small factories (ibid., 239).

Berg's study also called into question the accuracy of the statistics produced by Crafts and Harley. Thus she pointed out that their data excluded much of the vibrant small-scale production. Berg also argued convincingly that the whole gamut of production undertaken largely by women was also missing from the statistical base. "In current accounts of the slow productivity growth of British industry . . . traditional, male labor surplus industries take the fore while the new manufacturing industries deploying cheap female labor in conjunction with improved forms of economic organization and new machinery are excluded, or their place is understated" (ibid., 4). Although the picture of England's early industrial experience has changed once again, this new research, supplemented by other recent publications (Hudson 1992; 1996), has revived much of the older image of a dramatic "upward turn" in the English economy of the late 1700s and early 1800s.

Historians have also returned to the idea, emphasized in all the standard works on the Industrial Revolution, that foreign markets and supplies played a critical role in the Industrial Revolution. Although the new statistics unearthed by Crafts did not reveal as sudden an increase in the volume of eighteenth-century exports, a newer series by Javier Cuenca Esteban showed the percentage of industrial output sent abroad rising steadily from 20 percent in 1700 to 40 percent in 1801. "Overseas demand in general provided the opportunity *and the stimulus* for technological innovation as the industry reached the limits of growth within a projected domestic market," writes Cuenca Esteban (1997, 16). Furthermore, the nature of these exports was changing over to products tied into the new industrial processes and thus spawned greater efficiency, specialization, and eventually new technologies. Thus the significance of trade for early English industrialization, writes Patrick O'Brien (1993), was "not as expendable as [the quantitative historians] suggest." Finally, a significant new work by Kenneth Pomeranz (2000) demonstrates that access to raw materials in the American and Caribbean colonies may have provided the necessary boost to something like industrial takeoff. His main argument revolves around the importance of cost-effective slave labor in procuring for England vast quantities of sugar, timber, and cotton that freed up constraints on domestic land use. In order to produce the equivalent amounts of these materials or substitutes (e.g., wool and flax for cotton) at home, a massive amount of real estate would have been necessary—altogether about 11.5 million acres in 1811. This much more intensive land use would have limited food supply, restricted population growth, and drawn labor away from industrial ventures. In short, the Industrial Revolution might well have been choked off—even with cheap supplies of coal coming into greater use—just as it had been in China for these same reasons.

Recently economic historians have also challenged the idea that science played a significant role in early industrialization. In the works of A. E. Musson and Eric Robinson (1969), Margaret Jacob (1981, 1988, 1997), and Larry Stewart (1992), a consensus had begun to emerge that technical change in eighteenth-century England owed much more to science than was previously thought. The traditional view in works by Ashton (1955), Mantoux (1961), and even to some extent Landes (1969) held that England's technological breakthroughs resulted from "hands-on" trial-and-error experimentation by talented individuals who had no formal scientific knowledge. Economic historians also relegated science to the shadows, attempting to explain the great transformation as the inevitable outcome of market forces (Mathias 1969). This essentially skeptical assessment of science's contribution to industrialism has now surfaced again in the work of Joel Mokyr (2002). In a direct criticism of Jacob, in particular, he writes:

> Most practical useful knowledge in the eighteenth century was unsystematic and informal, often uncodified and passed on verti-

> cally from master to apprentice or horizontally between agents. Engineers, mechanics, chemists, physicians, instrument makers, and others could rely increasingly on facts and explanations from written texts, yet the instinctive sense of what works and what does not remained a critical component of what was "known." . . . Depending on what one means by "understand," it is obvious that people *can* do things they do not understand, such as build machines and design techniques on the basis of principles and laws that are poorly understood or misunderstood at the time (ibid., 30).

We should certainly not ignore such arguments. The metallurgical innovations just described, for instance, remain the best example of "hands-on" technological artistry during the first Industrial Revolution. It would take another century before scientists could even begin to explain the complex chemical reactions of the iron furnace. But it bears repeating that technical solutions do not appear like a genie when society rubs its magic lamp—either the "epistemic" (i.e., knowledge) base had to be wide enough in eighteenth-century England to support such solutions, or, as Mokyr would have it, technicians and entrepreneurs benefited less from scientific knowledge than from "scientific culture," one that not only familiarized them with methods of scientific experimentation but also "laid the intellectual foundations of the Industrial Revolution by providing the tacit and implicit assumptions on which technological creativity depended" and allowed those connected to the process of technological change "[to move] effortlessly between experimental science and practical applications" (1999, 77). The distance between Jacob and Mokyr may not be all that great, in other words. Whether one agrees with Mokyr that scientific understanding and insights—the actual epistemic base—was still quite primitive, or with Jacob that it was more sophisticated, a synthesis could once again be emerging that science one way or another played a key role.

Indeed many of England's path-breaking inventions were products of the scientific Enlightenment. Both Jacob (1988) and Stewart (1992) have demonstrated direct connections between popularized Newtonian science and technological innovation in eighteenth-century England. The latter writes with particular reference to steam engines in the mines, for instance:

> Natural philosophy and commerce were locked together in the embrace of the incipient industrial ventures of Hanoverian England. Natural philosophers—indeed, all those laying claim to special technical knowledge—were much sought after by investors and improvers throughout Britain and parts of the Continent. This cannot be surprising. . . . A superficial view of the eighteenth century might suggest few links between science and technology. Yet so many of the disciples of Newton were engaged in projects that the role of the natural philosopher underwent a profound transformation (Stewart 1992, 361, 391).

Other evidence abounds. Roebuck, whose lead-chamber process revolutionized bleaching, knew from university training in medicine that lead resisted sulfuric acid. John Smeaton and James Watt, the pioneers of England's steam technology, were well connected with the network of societies and universities that spread the new science. "The special thing about Watt's achievement," writes a recent historian of science, is that he took "a set of ideas from the cutting edge of then-current research in science and [applied] them to make a major technological advance; and the fact that he was working at a university, in direct contact with the researchers making scientific breakthroughs, presages the way in which modern high-tech industries have laboratories with close research connections" (Gribbin 2002, 249). And Matthew Boulton, the entrepreneur who financed Watt's work, possessed many of the same connections to the world of science. Some level of knowledge of physics and mechanics was absorbed not just by carpenters and millwrights, therefore, but also by merchant-manufacturers and proto-industrialists—producers who faced risky investment decisions as they advanced the technical transformation of eighteenth-century England.

EUROPE MARCHES IN PLACE

All these English developments stood in stark contrast to the situation in Continental Europe. Indeed the shortage of practical scientific knowledge as it applied to machinery was one of the major factors inhibiting an English-style industrial revolution there in the late 1700s. In order to understand why, we must return to our discussion of scientific trends in the previous century. Magical and mechanical approaches to science, both of which held promise for industry, competed with an Aristotelian establishment that was hostile to change—and remarkably resilient. Thus the Academy of Experiment in Florence, founded in 1657 to propagate Galileo's ideas, became a victim of the Catholic Inquisition a short ten years later. In France, similarly, Aristotelians prevented the adoption of René Descartes' mechanical views in three-fifths of the nation's four hundred liberal arts colleges by the early 1700s. Cartesianism spread more quickly and easily in Holland and the Protestant parts of Germany, but even here its acceptance was sometimes blocked by conservative clergymen who were worried, as Newton had been, about the potential for atheism, pantheism, and political radicalism (see above) (Armytage 1965; Jacob 1997).

Descartes wanted his new science to aid men of practical affairs. The brilliant Frenchman's *Discourse on Method* (1637) was intended for artisans, manufacturers, and merchants "who ordinarily have not studied at all." Descartes believed that reason and knowledge, once applied "to all the uses for which they are appropriate," could "lessen man's work" and make men "masters and possessors of nature." Despite these utilitarian professions, however, Cartesianism was not an ideal science for industrialists. Its approach to problem solving was far too theoretical, deductive, and erudite to benefit merchant investors who required demonstrations and applications of scientific theories to real engineering problems.

Propagators of Newton's natural philosophy, such as Jean Desaguliers, managed to take Newtonianism in this tutorial direction. But their efforts to transplant his system to Europe, while not entirely unsuccessful, met determined resistance from suspicious academics who either adhered to Aristotle or were devoted to Descartes. Once accepted at the University of Louvain (Austrian Netherlands) in the 1670s, for instance, Cartesian thinking became entrenched there until far into the eighteenth century. Dutch scientists at the University of Leiden proved more amenable to Newton's system. "It would flourish even more," lamented one true believer, "but for the resistance of certain prejudiced and casuistical theologians." Protestant Freemasons attempting to spread Newtonian science found kindred spirits at the British-oriented University of Göttingen in Hanover, but they were arrested in Paris and sometimes tortured in Italy and Portugal. The pope condemned Freemasonry in 1738. The effect of these cultural struggles across most of Europe, noted Joseph von Sonnenfels, an Austrian reformer of the late 1700s, was a "neglected" science that lagged a century behind England. "The evidence ... strongly suggests," concludes Margaret Jacob, "that many of the very men who had access to capital, cheap labor, water, and even steam power could not have industrialized had they wanted to: they simply could not have understood the mechanical principles necessary to implement a sophisticated assault on the hand manufacturing process" (1988, 181, 185).

Jacob's implication that most eighteenth-century manufacturers in Europe felt little urgency to advance beyond the existing technological system is important. Until catching up with industrializing England assumed a higher priority during the first half of the nineteenth century, there was simply less economic pressure on continental producers to mechanize and modernize their operations. Let us consider, for instance, the situation within the hundreds of semi-sovereign states that made up the Holy Roman Empire of the German Nation. Stretching from Austria and Bohemia in the Southeast, over Saxony, the Brandenburg plain, and the Hanseatic towns in the North, to the Rhineland and Belgium in the West, the lands of the empire never entirely recovered from the devastation of the Thirty Years War (1618-1648). German Europe lost over a third of its population during the conflagration and took a century to fully recover. In the war's aftermath, a technological system that had approached its limits now seemed more than adequate. Manufacturers in desperate, shrunken towns like Nürnberg, Augsburg, and Aachen tightened guild controls over the old methods of production in a vain attempt to protect shriveled markets. The fighting also left mines and metallurgical centers that had once been the envy of all Europe wrecked, flooded, or idle. In awe of their predecessors' knowledge, Austrian, Saxon, and Prussian mining officials were proud to have restored some of these sites by the late eighteenth century. "In technical matters we must adhere to the old ways," said one of them (Brose 1993, 136; Holborn 1964).

German population surged 30–50 percent above pre-1618 levels in the second half of the eighteenth century. For a variety of reasons, however, this impressive growth did not stimulate industrialization. First, it was not accompanied by an English-style export explosion. While Austria and Prussia, the mightiest German states, exchanged damaging blows with one another over control of Silesia, the great competition for world empire accelerated to the advantage of Britain and—to a lesser extent—France. Furthermore, it was Britain, not Germany, that enjoyed relatively fewer constraints on land use as a result of massive imports of raw materials from the new world. Making matters worse, both Britain and France penetrated German markets, claiming (with other countries) about 10 percent of all sales of manufactured goods. Domestic output also rose as population increased, but the miserable standard of living of the lower classes in central Europe minimized purchasing power and depressed demand. Cottage industry accounted for most of the increased output, as merchant-manufacturers escaped guild restrictions and expanded operations into the countryside. The putting-out system accounted for 43 percent of German manufacturing by the late 1700s (Gagliardo 1991). However, somewhat like England, these cottage industries coexisted with artisanal cooperatives and centralized operations employing a division of labor, oftentimes within different productive stages of the same product (e.g., spinning, weaving, finishing). Unlike England, however, institutional conservatism and technological skepticism on the part of merchants, guild leaders, and suspicious state officials stifled innovation (Ogilvie 1996).

In some regions, villages and small towns near guild cities did manage to grow into impressive centers of production. Elberfeld, Barmen, and Krefeld were good examples of this new type of textile town. In Monschau, near Aachen, entrepreneurs approached the threshold of industrialization by reassembling yarn in town for weaving in factories. In the nearby Bishopric of Liège, merchants had even advanced to the point of considering the installation of textile machinery. Yet, not surprisingly, this mechanical technology proved beyond their ken, prompting determined agitation by businessmen and Freemasons for scientific-technical education. None of this moved the bishop, who would sooner have supped with the devil than yield to the new, "subversive" science (Holborn 1964; Jacob 1997).

France offers perhaps the most interesting comparison with England, for, as noted above, it was the island nation's greatest imperial rival. It may surprise many readers that despite its military defeats to England during the eighteenth century, France more than held its own economically. Its overall manufacturing growth equaled Britain's, while population and value of foreign trade actually rose slightly faster. The basic and glaring difference between the two economies was technological: England began to invent new ways to produce goods and France, with some notable exceptions, did not. Its pig-iron output rose impressively to about 135,000 tons, but iron makers retained charcoal as fuel. Its textile industries kept pace with England, but, when organized in factories, they were powered by hand or water, not steam. And cottage industry accounted for more of this textile expansion than did the larger-scale factory operations.

Economic historians like François Crouzet argue that France's expansion within the technological system that had evolved in the sixteenth and seventeenth centuries resulted from material factors. Put simply: wood and labor were abundant and cheap, while coal was scarce and expensive. To these "supply-side" explanations, however, we should add a contrasting scientific culture and a dearth of English-style scientific and technological knowledge. As Crouzet himself admits, French manufacturers "were quite ready to take up foreign inventions (although they often had difficulty making them work)." And although France "had plenty of scientists . . . all this intellectual activity had very little practical result; discussions in learned societies remained theoretical in character and those who took part lacked a sense of the concrete" (Crouzet 1967, 159, 160-61). France suffered, in short, from the erudite legacy of Cartesianism as well as a sociopolitical division (and attendant lack of communication) between scientists entrenched semiofficially in the Academy of Science on the one hand and businessmen on the other. Jacob's (1997) discussion of the frustrations and delays experienced by entrepreneurs who attempted to import Watt steam engines in the 1770s, only to experience the haughty and indifferent superiority complex of academic scientists, is an excellent example. Taken together with the availability of wood, the abundance of rural labor, and the shortage of coal, these social, intellectual, and political factors help to explain France's technological lag in the eighteenth century.

France's new leaders experienced the resiliency of these material and nonmaterial forces with mounting frustration in the quarter century after 1789. With one notable exception (see below), the institutional setting could not have been much better for a synergistic spiraling of science and technology. The absolute monarchy yielded to a succession of regimes that allowed the middle classes access to the governmental machinery—even under military despotism after 1799. Indeed Napoleon's establishment of the Bank of France in 1800 epitomized the openness of the new order. The old conservative colleges were abolished and replaced with new schools and institutes committed to scientific and technical education. The government awarded lucrative prizes for the emulation of British technology and founded a special society to encourage technological advancement. The nation also created a huge continental market for its industries with unparalleled conquests, high tariff walls, and a continental "blockade" of English goods. Yet when the wars ended in 1815, French industry had registered little progress. Iron makers had expanded along traditional lines, rejecting coke fuel on grounds of quality and expense. War-induced labor shortages accelerated mechanization, but in the one area where France threatened to steal a technological march on England—machine-tooled guns—army conservatives interfered. Preferring the social benefits of artisanal production over the technical and economic advantages of mechanization, the war ministry cut off government contracts to the innovative firm of Honore Blanc and thus squelched the embryonic process of "interchangeable" parts. By 1815, the techniques had been forgotten—a form of "technological amnesia," according to one recent historian (Alder 1997, 319). Consequently, the English were even further ahead than before 1789, and the French still lacked the practical mechanical knowledge and hands-on experience to close this widening gap. The insatiable demand in France (and other parts of Europe) for British mechanics and engineers—over two thousand had been lured across the channel by 1825—is the best measure of this deficiency.

THE TECHNOLOGICAL SYSTEM OF THE FIRST INDUSTRIAL REVOLUTION

Meanwhile, England accelerated into the industrial age. A few key statistics tell the story most clearly and succinctly. The output of woolen cloth, a traditional product, rose from 90 million pounds in the 1790s to 128 million in the 1840s. Cotton rocketed past its rival, however, soaring from 28.6 million pounds of raw material consumed in the 1790s to 550 million in the 1840s. Sulfuric acid tonnage increased from barely a thousand in the late 1700s to around three hundred thousand in the 1840s. Coal production, which stood at 6 million tons in 1800, grew to 60 million in 1850. Pig iron raced upward from 258,000 tons in 1806 to about 2 million in 1847. Steam-engine horsepower, including mines and transportation, leaped from 57,000 in 1800 to 620,000 in 1840 (Forbes 1957; Hyde 1977; Landes 1969; Mathias 1969; Musson and Robinson 1969; Paulinyi 1991).

It will come as no surprise that this marvelous half century also witnessed tremendous technological progress. In cotton textiles, for example, mule spinning came into its own as the machine grew in complexity, speed, and automaticity. By the 1830s, "self-acting" (i.e., automatic) mules of twelve hundred spindles tended by one factory

Science and Technology in the World's Workshop 67

Figure 2.7: Mule spinning. Notice linkages to overhead pulleys and shafts connected to the steam engine. Courtesy of London Library.

operative performed ten times the work of their turn-of-the-century ancestors (see figure 2.7). Mechanization also spread to the spinning of wool. Jennies and simpler mules appeared in the West Riding area in the 1780s and 1790s, yielding gradually to semi-automatic mules by the 1830s and 1840s. These machines were still more common for worsted cloth, whose combed fibers withstood more strain than regular woolens, and their operating speeds were only about two-thirds as fast as the machines used for cotton. Self-acting machines for spinning wool, moreover, were a rarity until after 1850. It was mandatory, of course, that weaving keep pace with spinning. The "power loom" for cotton yarn, patented in 1822 by Richard Roberts, finally solved earlier breakage problems. Accordingly, the number of mechanized looms rose from twenty-four hundred in 1813 to a hundred thousand in 1833. Working at slower speeds, the power loom spread to worsteds in the late 1830s and woolens in the late 1840s (Landes 1969; Mann 1958; Paulinyi 1991; Usher 1967).

Developments in textiles challenged Britain's chemical industry to better its performance. A significant breakthrough resulted from the production of bleaching powder by Charles Tennant of Darnley, Renfrewshire, in 1799. Although preceded by the scientific discovery of chlorine in Sweden (1774) and the discovery of its bleaching properties in France (1785), chlorine found industrial application in the burgeoning cotton-textile industries of England. Tennant applied sulfuric acid to common salt and manganese dioxide in a lead-lined chamber, passed the mixture through a second lead chamber containing water, then combined this with lime in a third stone chamber. After two days, the lime mixture was stirred with rakes and allowed to dry. "But for this bleaching process," commented one expert, "it would scarcely have been possible for cotton manufacture in Great Britain to have attained the enormous extent which it did during the [early] nineteenth century" (Clow and Clow 1958, 248). The rise of dry chlorine bleach made it necessary, in turn, to economize the production of sulfuric acid. Again, Tenant improved the lead-chamber process around 1810 by burning sulfur and nitrates more efficiently in a separate furnace and replacing water on the floor of the vats with jets of steam. Not only did the furnace gases combine much more rapidly with steam, but this procedure also produced a more concentrated solution of sulfuric acid (Paulinyi 1991).

Technological change in the British iron industry was no less impressive. Blast furnaces grew much taller, for instance, finally reaching heights of eighty feet that permitted greater efficiency and economies of scale. Iron makers also devised techniques to improve the quality of coke iron. Thus furnace masters realized through trial and error that lime and calcium had a powerful affinity for sulfur. Hence, limestone used in excess of the amount normally used as a flux improved quality by reducing sulfur content. The so-called hot-air blast delivered to the iron furnace from an adjacent oven also reduced impurities by raising furnace temperatures. At six hundred degrees Fahrenheit limestone combined even more readily with sulfur; less coke was consumed per ton of iron ore; and therefore proportionately less sulfur was introduced into the furnace. The puddling process also saw improvements as technicians experimented with different materials in furnace construction, like an oxygen-rich slag to line walls, which combined quickly with carbon in the molten pig iron, sped the reduction process to wrought iron, and thereby shortened the time that sulfurous gases from the reverbatory furnace could contaminate the iron. Coke iron still lagged qualitatively behind charcoal iron in 1850, but quality differentials were narrowing quickly (Brose 1985; Hyde 1977).

The rapid substitution of iron for wood and other materials helps to explain the eight-fold increase in British iron production. The strength and durability of bridges, buildings, machines, and other structures was tremendously enhanced as a result. This transition would not have occurred, however, without the simultaneous emergence of modern machine tools that could create the same shapes and make the same holes, slots, and grooves in iron as in wood, but with much greater precision. Two of the earliest and most important were (1) boring machines (1770s) that cut steam-engine cylinders exactly enough to allow for the movement of pistons with no release of steam and (2) rolling mills for flattening red-hot puddled iron into bars (1780s). Lathes for making screws came into use around 1800; ingenious metal-planing machines appeared in the 1810s, while shapers for grooving and milling machines for cutting gears and nut heads came on the scene in the 1830s. Industrialists installed one of the largest machine tools, a steam hammer to handle the bigger iron forgings of the burgeoning industrial era, in 1839. Although less dramatic, a punching machine (1847) for placing bolt holes at regular intervals increased productivity just as demand rose for iron plate used in bridges, boats, and steam engine boilers (Ferguson 1967a; Gilbert 1958; Paulinyi 1991; Pursell 1997).

Steam engines were another maturing technology of the half century after 1800. To be sure, Watt's basic design of ten to twenty horsepower was adequate for most factory uses. By 1840 these kinds of engines had relegated waterwheels to an auxiliary role, generating 72.5 percent of horsepower, for example, in Britain's textile factories. The only significant improvement to the Watt engine came in 1845 when John McNaught added a second "high-pressure" cylinder on the beam near the crankshaft, thereby boosting horsepower to about sixty. The "compound" engine and the high-pressure boiler were not new, however, for they had found application in the mines of Cornwall after the mid-1810s. Very dangerous, the latter design increased steam pressure per square inch from the two or three pounds above atmospheric pressure of Watt's engines to between thirty and fifty—and well over a hundred pounds in some instances. High-pressure engines were a significant breakthrough, for they produced great force relative to engine size and weight and could therefore power steamboats and railroad locomotives (Dickinson 1958; Ferguson 1967b, 1967c; Musson and Robinson 1969).

The railway age, which unfolded after the mid-1820s, was a product of many technologies. Mines burrowed deeper and wider into the earth to provide millions of tons of ore and coal for furnaces and rolling mills that turned out iron for sixty-six hundred miles of track between 1825 and 1850. Expensive tunnels, bridges, and viaducts utilizing techniques borrowed from road and canal construction kept the lines straight, level, and cheap to operate. Machine-tool plants cut, bored, planed, and punched iron into screws, bolts, clamps, plates, and the myriad metal products needed to fasten track to ties, secure bridges, build station houses, and produce railway cars and locomotives. Most impressive of all, perhaps, were the great smoking colossuses that pulled passengers and products (see figure 2.8). The prototypes debuted at the Rainhill Trials of the Manchester-Liverpool Railroad in 1829. George Stephenson's "Rocket" won the competition, easily hauling its twenty-ton load up moderate gradients at speeds of fifteen to thirty miles per hour. Its multiple-fire-tube boiler generated more steam and power than the older, single-flue types by doubling the area of heating space relative to engine weight. By 1832, Stephenson's locomotives possessed even better boilers that could withstand pressures of fifty pounds per square inch. With a third set of wheels to distribute the engine's seven tons, this design generated twice as much tractive power as the original Rocket. Railroads now came into their own (Ferguson 1967c; Paulinyi 1991; Wescott 1958).

The complex technological interactions of the railroads are a reminder that England was shaping a new technical system. Its evolution and maturation are illustrated by the flow chart in figure 2.9. Sulfuric acid, iron, and the steam engine formed the integral component parts of this technological system. By 1850 the steam engine, a "technical ensemble," that is, an assemblage of specific techniques into one machine or process, had linkages to other ensembles like machine tools, steamboats and railroads, steam plows, and printing presses as well as to the mining-metallurgical and textile "concatenations," that is, vertically linked ensembles, each responsible for one stage of making a final product—the spinning, weaving, bleaching, and finishing of cotton cloth, for example, or the mining of iron ore, its conversion to pigs and then to wrought iron that could be shaped and hammered for various uses. This crucial technical ensemble literally powered the Industrial Revolution, enabling Great Britain to achieve levels of output that would have been impossible given the limitations of water, wind, and animal power. Cheap iron of passably good quality was another prerequisite of industrialization, for wood was an inferior material unsuitable for most purposes of the new era. Accordingly, cast- and wrought-iron use proliferated from 1750 to 1850, extending to machine frames, machine parts, and machine

72 Technology and Science in the Industrializing Nations, 1500–1914

Figure 2.8: English steam locomotive of 1813 (*top*) and German locomotive of 1840 (*bottom*). Courtesy of Science Museum, London; Hans-Josef Joest, *Pionier im Ruhrrevier* (1982), p. 25.

tools; steam engines; steamboats; track, bridges, and locomotives; pipes, beams, and construction products; plows and farm implements; as well as pots, pans, utensils, and other household items. The lead-chamber sulfuric acid process was another technical ensemble that supported a superstructure of numerous ensembles in many concatenations. Indeed by the early 1800s manufacturers used this crucial acid for bleaching, printing and dyeing, match making, paper making, fertilizer, soda making, and through soda to soap, glass, and saltpeter. The central role of steam power, iron materials, and sulfuric acid in so many industrial sectors; the many interactions between ensembles and concatenations; and the mutual technological dependencies present throughout the industrial economy lent a systematic character to the nation's technical processes. It should come as no surprise that England experienced a noticeable increase in production as this technical system came into its own—even the sober studies of Crafts and Harley (1992) show industrial growth rising to the 3 percent range in the 1820s and 1830s.

Quantitatively, Great Britain had surpassed all other European nations. The island power turned out four times as much pig iron as France (1847), for instance, and English steam-engine capacity soared three times above that of France, Belgium, Germany, Austria, and Italy combined (1840). As late as 1870 continental Europe (including Russia) produced barely 200,000 tons of sulfuric acid to England's 590,000. By this time Great Britain had become the "workshop of the world."

More telling than these numerical measurements, however, was the fact that Europe's technological system was only beginning to transform itself from Early Modern times. Railroad construction accelerated in France and Germany during the 1840s—and here and there modern textile, machine-tool, and chemical plants sprouted up—but slower-moving wooden machines powered by water stayed the norm, and charcoal iron remained king. In short, while Europe may have modernized some of its operations before mid-century, only England had devised a new technological system from one end of the industrial sector to the other.

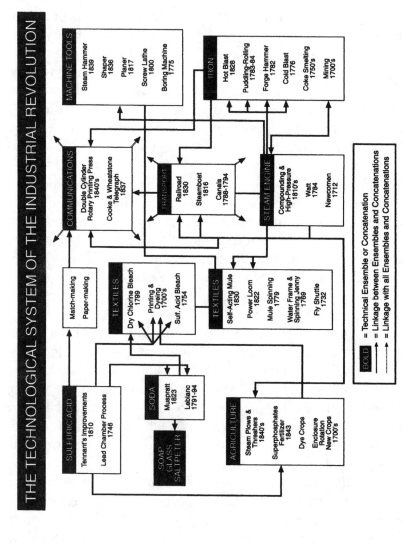

Figure 2.9: Flowchart of the technological system of the English Industrial Revolution of the mid-1800s.

* * *

The technological system of the first Industrial Revolution brought tremendous economic, military, and political advantages to England that made it the envy of Europe. In the decades after 1815, however, continental countries attempted to convert this envy into a successful struggle to narrow the gap. In the process, an altogether new scientific-technological system based primarily in Germany—and, in modified form, the United States of America—emerged on the scene. These developments are the subject of our final chapter.

III

LANDS OF UNLIMITED POSSIBILITIES

On a gray day in January 1911, the elites of the mighty German Empire gathered in Berlin. The occasion was the inaugural meeting of the Kaiser Wilhelm Society for the Advancement of the Sciences. Representing the social forces of an older, resilient order was Emperor Wilhelm II himself, resplendently dressed in a uniform appropriately befitting the military implications of civilian science in the modern era. All around him were the giants of Teutonic ingenuity: the heads of the nation's dominant dye, pharmaceutical, and explosives firms and the most prestigious professors of physics and chemistry from the university establishment. Old Germany and New sat somewhat uneasily and uncomfortably together in the minutes before the keynote address.

To the podium stepped a bearded, bespectacled, distinguished-looking man in formal attire. It was Emil Fischer, a professor at the University of Berlin, member of the prestigious Academy of Science, winner of the Nobel Prize, and undoubtedly the world's leading organic chemist. Born in 1852, Fischer had risen to the head of his field by the age of forty. Along the way he had done more than any other academician to foster cooperation between academe and industry, thereby contributing singularly to Germany's position atop the industrial world of the early twentieth century. But it was a very competitive industrial world. England had still to be reckoned with, while across the Atlantic, the United States of America had transformed itself from an agricultural to an industrialized nation. It combined the advantages of vast natural resources; a rapidly growing, hard-working population; and, more recently, university-based scientific research liberally funded by private industry. Ludwig Goldberger, a businessman present in the audience, had pointed with alarm to Germany's transatlantic competition in 1903, dubbing America "the land of unlimited possibilities." Similar concerns had prompted Fischer and his colleagues a few years later to propose the Imperial Institute for Chemistry, which would accelerate research in Germany with government support. But cost-conscious bureaucrats had scuttled the proposal. By tapping into private donations with the power of the emperor's name, the Kaiser Wilhelm Society was designed to reverse these recent failures.

Germany's threatened preeminence in the world weighed heavily on Emil Fischer's mind as he turned to face his colleagues and his sovereign. Although the empire's scientists had won twelve Nobel prizes in the first decade of the award—six of them coming in chemistry—past greatness could not assure future success. Today's scientists were finding it particularly difficult to combine burdensome teaching loads and ambitious research agendas. Only heavily funded institutes devoted exclusively to research could free Germany's young minds, meet the American challenge, and realize the tremendous potential of science. Fischer dangled an exciting list of recent or potential breakthroughs before his audience: radium substitutes, synthetic food and drink, liquefied hydrogen, artificial rubber, and fertilizers and explosives produced from atmospheric nitrogen. As a military-minded Wilhelm nodded in apparent approval, Fischer claimed that these products of the chemical lab, not inaccessible overseas colonies, were the "hidden treasures" that would enable Germany to surmount foreign threats. The Kaiser Wilhelm Society would enable German scientists, the "Kaiser's godchildren," to chart their way through chemistry to a "true land of unlimited possibilities" (Johnson 1990, 126–27).

GERMANY AND THE SECOND INDUSTRIAL REVOLUTION

One hundred years earlier, as industrialism accelerated in England, few continental observers had reacted positively to the unleashing of boundless potential across the channel. Indeed the notion that Europeans should emulate scientific and technological developments in England was anathema to all but the boldest and most enterprising merchants and businessmen. Led by French and German administrators who had cast envious glances at England since the late eighteenth century, the bureaucrats tended to be somewhat more progressive, although even "pro-industrial" states like Prussia often put more obstacles in industry's path than they did to remove these barriers (Brose 1993). This insightful minority in and out of government watched with alarm as inexpensive chemical products spilled into European markets after the Napoleonic Wars. Tariff walls offered some protection from this onslaught, but smugglers usually found an opening through these porous barriers. Therefore, unless their products were qualitatively superior to Britain's factory-made goods, or unless high transportation costs to the remotest parts of Europe insulated them from competition, the wisest producers realized that survival depended exclusively on the adoption of British productive techniques. Because such a course of action was fraught with difficulty and risk to investors, however, holders of wealth asked governments to help.

Yet discussion of this policy option sparked even more controversy than did English dumping. Thus the guilds, those medieval corporations of privileged handicraft manufacturers, argued that it was wrong for bureaucracies to allow new businesses to produce in new ways, for this would violate the guilds' centuries-old right to limit and regulate output. Governments should see the novel English methods for what they were: a devilish plot to ruin families and spread starvation. Social observers of both liberal and conservative persuasion tended to agree with the artisans. Travelers to the smoky industrial island reported the growth of an impoverished working class congregated in ugly, dirty, riotous towns. Did it not make more sense to reinforce the old social order than to sacrifice it for such misery? Soldiers were also wary of industrialism. Long working hours in unsafe factories weakened body and soul and fighting spirit. To critics of England's Industrial Revolution, the state's role should be to stop the technological wheel from turning by raising tariffs, catching smugglers, enforcing guild privileges, and banning factories (Brose 1997; Jacob 1997).

The rejoinder from the other side was that economic disaster and national decline awaited countries that ignored the gauntlet thrown down by England. A successful response to this challenge, on the other hand, would preserve jobs, guarantee taxes, and line government coffers. States should improve transportation, provide tax breaks to aspiring factory owners, suppress the guilds, teach modern subjects in school, and help businessmen access the new technology. Thus the industrialization of nineteenth-century Europe became a matter of political choice.

In German Europe this debate climaxed between the 1810s and the 1830s. Gradually, the Cassandras, who predicted that social and economic troubles would result from the coming of industry, were silenced by opponents who made their cases more persuasively. The outcome of this political struggle was not one-sided as governmental factions in Berlin and Vienna continued to frustrate businessmen with restrictive regulations, but gradually another set of policies with the explicit aim of promoting industrialization predominated. In Austria, for instance, enlightened bureaucrats eliminated internal tolls and raised external tariffs to prohibitive levels. By preserving the lion's share of the empire's huge market for domestic sales—smugglers would take their part—civil servants hoped that industrialists would be emboldened to invest in large-scale ventures that utilized the latest techniques. To facilitate the adoption of state-of-the-art technology, the state assembled foreign machinery in a permanent exhibit in Vienna—some twenty thousand pieces by 1822. Prussia, Austria's rival in northern Germany, adopted similar practices. Internal free trade combined with much lower external tariffs would force factory owners to sink or swim with the British competition. Bureaucrats in Berlin followed Vienna's lead by gathering a large collection of models, blueprints, and samples of recent machines from England. Prussia's construction of new roads was also particularly impressive, totaling thirteen thousand kilometers (8,078 miles) from 1816 to 1850. Although some German states resisted the temptation to emulate England—witness the anti-industrial policies in Bavaria, Hesse-Kassel, and many of the smaller city-states—the overall trend was determined by the two largest powers. This was especially true with the advent of railroads in the 1830s. After initially opposing railroads, both Austria and Prussia promoted the new mode of transportation after the mid-1830s, accounting for 89 percent of German Europe's four thousand kilometers (2,486 miles) of line by 1847. The pace of industrialization quickened.

As we emphasized in chapter 2, however, technological development is never an automatic response to societal demands and economic needs. In order for a society to generate new forces of production—or, in Europe's case, to catch up and overtake the world's technological leader—the appropriate scientific culture must be available and accessible. The mechanical revolution of the eighteenth century sprang from the gradual permeation throughout England of applied Newtonian physics. With neither the economic incentive nor the appropriate scientific preparation, continental emulation was at first out of the question. Then, with the stakes growing higher after 1815, Europe strove to acquire the knowledge behind England's revolution. Continental entrepreneurs hired British mechanics by the thousands and at great expense. Permanent machine exhibits and collections came into being. But the most informed European policy-makers knew that educational reforms were the only long-run strategy that promised success in acquiring and disseminating scientific knowledge.

And here central Europe led the way. The first school designed with the explicit purpose of spreading scientific and technological knowledge to businessmen and industrialists was established at Prague in 1806. A second institution of this sort opened its doors in Vienna nine years later. The basic curricular model for both institutes was France's famous École Polytechnique. Like their counterparts in Paris, students studied physics and mathematics. Beyond this the two curricula differed, however, for Vienna and Prague, educating future businessmen rather than the next generation of bureaucrats and army officers, offered additional classes in industrial chemistry, industrial engineering, and the history of industrial technology. Prussia followed suit in 1821. The most promising students in the kingdom were chosen from twenty-five regional technical schools for a further three years of technological training in the Business School of Berlin. Other German states now hurried to establish polytechnical institutes of their own: Baden in 1825, Saxony in 1826, Bavaria in 1827, Württemberg in 1829, Hanover in 1831, Brunswick in 1835, and Hesse-Darmstadt in 1836. "Whereas in former times men regarded the inquisition of nature as a pleasant but useless employment, and as a harmless pasture for idle heads," observed a German scientist in 1830, "they have of late years become daily more convinced of its influence upon the civilization and welfare of nations, and the leaders of the public are everywhere bestirring themselves for the erection of establishments to promote its advancement and extension" (Armytage 1965, 78–79). Altogether these institutes instructed about five thousand students annually by 1850 (Slokar 1914; Wehler 1987–95, vol. 2).

While most of German Europe's roughly forty thousand industrialists at mid-century were self-taught, having essentially "learned by doing," some of the most famous entrepreneurs and specialists were beginning to exit these schools. Out of Prussia's Business School, for instance, came August Borsig, the famous locomotive engineer and manufacturer, and Friedrich Wöhler and R. W. Bunsen, both illustrious chemists. The latter names remind us that Germany's universities, world-renown for their philosophers, philologists, and historians, were now also making a contribution to scientific understanding and technological transformation. Thus Wöhler built up an important chemical laboratory at Göttingen after 1836, teaching eight thousand students in twenty years, while Bunsen established his research facilities at Marburg in 1839, later moving to Heidelberg. The model for both men, however, was Justus Liebig's chemical research center at the University of Giessen. From 1825 until 1852, the father of German chemistry inspired students and followers to convert their scientific researches into useful knowledge. Meanwhile the polytechnical institutes expanded into full-fledged "technical colleges" (*technische Hochschulen*) with larger staffs, bigger enrollments, and higher professional status. As German industry expanded after mid-century (see below), an impressive scientific establishment was on hand to solve existing technological problems as well as open vistas onto that landscape of "unlimited possibilities" that would later enthuse Emil Fischer (Armytage 1965; Wehler 1987–95, vol. 3).

For decades many economic historians argued that German industrialization, fueled by many of the positive government programs mentioned above, accelerated in the middle of the nineteenth century. Alexander Gerschenkron (1994), to take one classic thesis, formulated a "big spurt" dating from the 1840s. According to the famous study of Walter W. Rostow (1960, 1963), moreover, German industries "took off " in the 1850s and 1860s—a model that may be more appropriate for Germany, in some respects at least, than England. Rostow posited a period of about twenty years when countries accelerated into the industrial era. A long period of preparation (characterized by institutional reform, acquisition of business skill, and investment in agriculture and infrastructure) yielded to decades of accelerating industrial growth self-sustained by strategically reinvested profits. Rostow also assumed that one sector of industry played a "leading" role, expanding so quickly that it took the rest of the economy in tow. In German Europe (i.e., Germany and Austria), railroads took this lead (Fremdling 1975), rising meteorically from 4,000 to 16,700 kilometers (2,486 to 10,377 miles) between 1847 and 1860 alone. During the same years pig-iron output expanded from 427,000 to 842,000 tons, while steam-engine capacity soared from 360,000 to 1,180,000 horsepower. The number of cotton spindles, which stood at 1.425 million in 1834, shot upward to 4.035 million by 1861 (Brose 1997). Although more recent quantitative and qualitative histories have shifted attention to the expansion of investment *before 1840* (Tilly 1994; Brose 1993), despite a mix of positive and negative governmental stimuli, there is also evidence that the post-1850 decades—Rostow's period of take-off and self-sustained, post-takeoff expansion—witnessed impressive annual industrial growth, fluctuating from 4.41 percent (1850–1875), to 2.62 percent (1874–1894), and then back up to 4.16 percent (1894–1914), with a noticeable speedup, 4.97 percent, on the eve of war (1908–1914). Rostow's assumptions about acceleration after takeoff clearly break down in the 1870s and 1880s, but German industries were still ascending despite the slowdown, and contemporaries and subsequent historians were not wrong to see it this way. As discussed below, furthermore, this growth certainly set the stage for the evolution, as Gille (1986) would see it, of a new and more dynamic technological system in the later decades of the nineteenth century. In other words, viewing German industrialization through the shifting lens of, first, Rostow and then Gille brings a complex phenomenon into better focus.

To be sure, England continued to enjoy a commanding quantitative lead during these critical years of mid-century. Its pig-iron output and number of cotton spindles were still more than three times that of Germany and Austria in 1870, and its sulfuric acid production was nearly triple the tonnage of the entire continent. Technologically, it was the British system that central European manufacturers strove to emulate. Self-acting mules and automatic looms, reciprocating steam engines, lead-chamber acid, coke pig iron, and precision machine tools made their way into the factories and furnaces of Rhineland-Westphalia, Silesia, Bohemia, and lower Austria.

As the decades passed from 1870 to 1900 and beyond, however, Germany began to close the gap as its industries grew more than twice as fast. One of the major stimuli for this impressive rate of growth was the formation of a united German Empire in 1871. Now there was only one currency, one national bank, one set of business and labor laws, and one powerful state to provide protection and security in Europe's explosive and violent atmosphere. By the turn of the century, Britain's output of pig iron, steel, and sulfuric acid actually fell behind that of Europe's rising power.

During these latter decades another significant development took place. For the technological system that had matured in England during the early nineteenth century, answering that nation's material challenges in spectacular fashion, started to reach its limits and break down. It was clear as early as the 1850s, for example, that malleable wrought iron could not remain the system's basic material. It sufficed for rails in the 1830s when locomotives weighed around a ton, but as engine weights approached twenty-five tons at mid-century—reaching a hundred tons by the early 1900s—iron rails wore out too quickly and replacement costs became a "financial abyss" (see figure 3.1). It appeared to be impossible to substitute harder and more durable steel, furthermore, for it was prohibitively expensive—twenty-five times as high as iron. The reciprocating steam engine also grew increasingly problematic. The accelerating demand for more power meant faster engine speeds, which, in turn, required more and more force to stop and reverse the momentum of the piston and connecting rod. The tremendous stresses of this back-and-forth motion necessitated frequent piston, cylinder, and rod repair and caused engine breakdown. Compounding difficulties, the alternating steam engine converted heat to energy at a very inefficient rate of 6 to 10 percent and was unable to transport power through overhead shafts and pulleys without a great loss of power (see figure 3.2). The steam engine of the first Industrial Revolution presented special difficulties onboard ship because the necessary increases in speed were unattainable without increasing power by approximately the cube of the change in speed (see figure 3.3). And larger engines meant larger engine rooms and coal bins as well as thicker hulls to withstand the shock. One study determined that optimum cargo space was achieved in ships of 29,500 tons, with the maximum load at about 51,000 tons—the size of the largest liners of the early 1900s. Available space would approach zero as ships grew to 95,000 tons. These examples of bottlenecks from materials, power, and transport were

Figure 3.1: Size comparison of the Rocket and a locomotive of the late 1800s. Courtesy of Corbis Inc./Bethmann Archive.

Lands of Unlimited Possibilities 87

Figure 3.2: German machine tool factory showing complicated and inefficient overhead linkages. Courtesy of Staatsbibliothek Preussischer Kulturbesitz, Bildarchiv, Berlin.

Figure 3.3: The great Corliss engine of 1876 demonstrating size and power constraints of reciprocating steam power. Courtesy of Metropolitan Museum of Art.

not isolated cases, for blockages occurred throughout the late 1800s as rapid growth forced the technological system to the limits of its potential (Gille 1986, 1:688; Landes 1969).

The result was the relatively quick development of a new system between the 1850s and the 1910s. "Even though certain old sectors, so to speak, were carried over into the new and more advanced technical system," notes Bertrand Gille, "one may observe that the techniques in use on the eve of the First World War were completely different from those which were generalized around 1850" (1986, 1:671–73). Indeed the coming of cheap, mass-produced steel, electrical power and equipment, the steam turbine, the internal combustion engine, the airplane, the telephone, precision-tooled interchangeable parts, synthetic dyes, and "contact" sulfuric acid created another industrial revolution characterized, like the first, by the "concomitance and convergence" of these techniques throughout the system. No single nation could claim these innovations as its own: their origin and widespread adoption were distributed among England, France, Germany, and the United States. As I now argue, however, the dynamism of the German economy and the brilliance and ingenuity of the empire's scientists and engineers lay behind much of this second great transformation.

The technical revolution in steel is a good example. Steel is essentially a form of iron with a moderate 1 or 2 percent carbon. This material is stronger than low-carbon (0 to 1 percent) wrought iron and more malleable and than high-carbon (4–5 percent) cast or pig iron. But its high price before the mid-nineteenth century—and consequent limited use—reflected the long, laborious process of heating, reheating, and decarburizing (i.e., reducing the carbon content of) pig iron. In the 1840s and 1850s, French and German iron makers modified the puddling process to produce steel by reducing furnace time and thus preventing a complete decarburization of the iron. Never very high in quality, "puddle steel" yielded during the 1860s to a conversion process devised simultaneously in England and the United States. The Englishman Henry Bessemer and the American William Kelly blasted hot air through molten pig iron, which reduced carbon content spectacularly in a few minutes (see figure 3.4). Unable to eliminate deleterious phosphorous from the finished steel, however, the process was restricted to pig iron smelted with expensive low-phosphorous iron ore until 1879. That year two Englishmen, Sidney Gilchrist Thomas and his cousin, Sidney Gilchrist, introduced a limestone slag to the converter while also lining the container with chemically "basic" dolomite bricks. Both slag and lining combined with phosphorous, removing this harmful element from the steel and permitting the use of more abundant phosphoric iron ores. A competing process developed in England by the brothers Frederick and William Siemens, naturalized Englishmen of German origin, and in France by Pierre Martin, employed heated coal gas to melt and decarburize pig and scrap iron slowly in a large hearth. Dating mainly from the 1870s, Siemens-Martin or "open hearth" steel was modified in the 1880s to accommodate the Gilchrist-Thomas process and utilize raw materials made from phosphoric ores. Iron struggled to compete but could not. Once so high, the price of steel had nearly dropped to that of iron by the late 1870s. For uses like rails, moreover, the newer material lasted six times as long. Consequently, steel output pulled even with, then passed, that of wrought iron—in England by 1885, Germany by 1887, and France by 1894.

Of the three great European countries, Germany pushed the revolution in steel forward with the greatest intensity. Not only had the empire invested less in the older methods than its western rivals, but also it required more material for its rapidly growing cities, completion of a larger rail network, maintenance of a greater army, and—after 1898—closing the naval gap with Great Britain. German steelmakers also benefited tremendously from the vast deposits of phosphoric iron ore in Lorraine, a western province taken from France as a spoil of war in 1871. On the eve of the Great War in 1914, Germany produced 9.7 million tons of Bessemer steel, while France made 3.3 million and England only 1.63 million. But the European powerhouse also turned out 6.5 million tons of open-hearth steel to England's 6.2 million and France's 1.69 million. In fact, German steel output was nearly equal to that of England, France, Italy, and Russia—the nations that would soon be in the field against Berlin.

Lands of Unlimited Possibilities 91

Figure 3.4: Bessemer converters in loading (*right*) and blasting (*left*) positions. Courtesy of Universitätsbibliothek der Technischen Universität, Berlin.

Germany's blast furnaces, Bessemer converters, and Siemens-Martin hearths were also larger and more efficient than anything in western Europe. Quantitatively and qualitatively, Germany was the king of European steel (Landes 1969; Schubert 1958).

We turn now to the new electrical industries. In contrast to metallurgy's largely empirical tradition, European science had gradually prepared the way for the commercial exploitation of electricity since the beginning of the century. "The landmarks," writes David Landes, "stand out: Volta's chemical battery in 1800; Oersted's discovery of electromagnetism in 1820; the statement of the law of the electric circuit by Ohm in 1827; the experiments of Arago, Faraday, and others, climaxed by Faraday's discovery of electromagnetic induction in 1831; the invention of the self-excited electromagnetic generator (Wilde, Varley, E. W. von Siemens, Wheatstone, et al.) in 1866–67; Z. T. Gramme's ring dynamo, the first commercially practical generator of direct current, in 1870; [and] the development of alternators and transformers for the production and conversion of high-voltage alternating current in the 1880s" (1969, 282). With the scientific groundwork prepared (also see below), electric power spread to a number of critical areas of the late-nineteenth-century economy. Private residential, commercial, and municipal lighting; streetcars and local railways; and fixed industrial motors were powered and driven increasingly with electricity by 1900 (see figure 3.5). As a power source it had the advantage (over steam engines) of easy starting and stopping without costly adjustments, quick and efficient transmission over long distances, as well as the flexibility and convenience of application to large factories, small businesses, or home workshops. "On the one hand, electricity freed the machine and the tool from the bondage of place; on the other, it made power ubiquitous and placed it within reach of everyone" (Jarvis 1958a, 1958b).

"In its relations to science, capital, and governments," observed John Clapham, "the electrical industry is a typical product of the last two decades of the nineteenth century—and in no other country can its characteristics be studied better than in Germany" (1966, 306). It was true. Germany's carefully cultivated scientific and vocational-technical schools, colleges, and universities placed the nation in a

Lands of Unlimited Possibilities 93

Figure 3.5: Electrical generator plant in Berlin, 1896. Courtesy of Berliner Kraft und Licht AG.

good position to contribute to and benefit from the scientific advances listed above. Imperial laws also favored the large business and banking concentrations and cartels so conducive to rapid growth in industries like electricity that required huge outlays and investments. By the end of the century the dominant firms were the massive General Electric Company of Emil Rathenau and the great Siemens-Schuckert combine (see figure 3.6). With 75 percent of the domestic market between them, the two giants grew at phenomenal annual rates: 9 percent in the 1890s and 16 percent after 1900. The positive results were evident by 1914. In terms of total electrical capacity and output, Germany had probably caught and passed Great Britain, the early pacesetter. But Germany was the undisputed world leader in the manufacture of electrical equipment and appliances. While its imports were negligible, imperial exports of cables, lightbulbs and lamps, motors, generating machinery, and other electrical products were more than two and a half times the British total and nearly three times the American. Thus the electrical crown also belonged to Germany (Hughes 1983; König 1990; Landes 1969; Wehler 1987–95, vol. 3).

The story line remains the same if we examine the burgeoning chemical industries of the late century. One of their most important functions in the technological system that propelled England to the top of the industrial world was in the finishing stages of textile production. Sulfuric acid and chlorine powder had broken the bleaching bottleneck, but dyes made from plants like indigo, madder (red), weld (yellow), cutch (brown), and moad (black and blue) soared in price as the escalating demand for colored fabrics pressed against the limited supply of land. The solution to this blockage was found in 1856 by William Henry Perkin, the young research assistant of one of Liebig's students, August Wilhelm Hoffmann, who had moved to England. Perkin accidentally discovered the first dyestuff derived from coal tar, a reddish color he called mauve. Many other coal-tar dyes appeared in England and France in the 1860s as a nascent industry sprang up to exploit the discovery (Holmyard 1966; Multhauf 1967).

By the following decade, however, momentum swung to Ger-

Lands of Unlimited Possibilities 95

Figure 3.6: German General Electric facility; generating station (*left*) and cable works (*right*). From Conrad Matschoss, *50 Jahre Berliner Elektrizitätswerke, 1884–1934* (n.d.), p. 30.

many as a result of its more systematic, scientific, and oligarchic approach to these problems. The rise of the German dye industry received impetus from the return of Hoffmann to the University of Berlin in 1865. In the capital he joined Adolf Baeyer and Carl Graebe, two of Bunsen's students in Heidelberg who taught at the Business Institute—formerly the Business School, and shortly to become the Berlin Technical College. The German ascent began with the creation of synthetic madder, or alizarin, in 1869 and culminated with synthetic indigo in the early 1890s. By 1914 German dyes manufacturers had been able to synthesize over a thousand different dyestuffs. The price of dyes fell to a third of its former level, and natural colorings were driven from the market. These synthetic products were produced by two giant cartels: one headed by the Badenese Aniline and Soda Factory (BASF), the other by the Hoechst Company. Each boasted laboratories that took invention and innovation from the day of ad hoc scientific engineering into the era of accelerated, large-scale, organizational or corporate research. Thus Hoechst had 165 chemists with university or technical college degrees; BASF employed 230 researchers; while the Bayer Company, Germany's third-largest dyes firm, gave full-time work to a phenomenal 600 academically trained chemists. The empire's chemical trusts also sold products aggressively throughout the world and serviced customers quickly and professionally. Small wonder that Germany dominated world dyes markets with a monopolistic 90 percent share (Beer 1959; Haber 1971; König 1990; Travis 1993; Wehler 1987–95, vol. 3).

Dyes were not the only product manufactured and exported successfully by the chemical companies of the German Empire. "Out of the lab research of the academic experts on synthetic dyes," writes Hans-Ulrich Wehler, "there developed an eminently significant 'spin-off' into entirely new and lucrative areas of production" (1987–95, 3:615). These included medical preparations, photographic film, artificial fibers, the first plastics, and powerful new explosives. More important than any of these, however, were two pathbreaking developments that spread throughout the technological system: a new and superior process for making sulfuric acid and a method of producing nitric acid from atmospheric nitrogen.

After spadework in England, the so-called contact process was patented in 1875 by Clemens Winckler, a professor at the Meiberg (Saxony), then brought to fruition in the 1880s by Rudolf Knietsch, a researcher at BASF. Under extremely high pressure and temperature, sulfur dioxide was burned in the presence of a platinum catalyst to produce sulfur trioxide gas, which then mixed in solution to produce sulfuric acid. The procedure required much less space than the bulky lead chambers, was much faster, and increased the acid concentration from 75 percent to 100 percent. With the better technology, Germany passed England, producing 1.7 million tons to Britain's 1.1 million in 1913. It was Fritz Haber, a chemistry professor at Karlsruhe subsidized by BASF, who produced ammonia from air in 1909. Haber borrowed from Winckler and Knietsch, developing this contact procedure that utilized an osmium catalyst to produce nitric oxide, which mixed in solution to make nitric acid. German science and industry had substituted an abundant source of nitrogen—air—for the rare and expensive nitrate ores used in the past to make this important acid. As I explain later, the contact technique was another interlocking component of the evolving technological system (König 1990; Landes 1969; Multhauf 1967).

The scientists, engineers, and businessmen of the German Empire contributed to this transition in other ways. The first practical internal combustion gasoline engine, for example, was the work of Nicolaus Otto in 1876, with improvements by Gottfried Daimler and Karl Benz in the 1880s (see figure 3.7). Aiding the growth of that technique, Germany's machine tools cut and shaped metal in state-of-the-art fashion (see figure 3.2). Without a doubt, however, the marquee technologies and leading sectors of the Second Industrial Revolution—Germany's technical triple crown—were steel, electricity, and chemicals.

TECHNOLOGICAL PROWESS IN FRANCE

It was France in 1850—not Belgium, Holland, any of the German states, or Austria—that seemed to be poised for a run at England's

Figure 3.7: Karl Benz at the wheel of his motorcar, 1887.
Courtesy of Deutsches Museum, Munich.

great industrial lead. At mid-century she mined more coal, smelted more iron, spun and wove more fabric, ran more factories with steam engines, and produced more sulfuric acid than any continental country. The railroad age had begun and growth rates were increasing. Indeed historians of the "takeoff" date this period of French industrial acceleration between 1830 and 1850. No other European nation had begun its takeoff, nor was any country as close to emulating Britain's system of convergent technologies. This was not surprising, for France had begun the trend in Europe toward more scientific and technological education with the establishment of the École Polytechnique in 1794. And France was still viewed by all Europeans as the greatest military power—a fact that seemed confirmed by its victory (with England) over Russia in the Crimean War (1853–56) (Clapham 1966; Dunham 1955; Landes 1969).

And yet by 1914 France had slipped to third place among Europe's industrial powers behind a resilient England and the dynamic German Empire. The image of relative decline is sharpest if one focuses on traditional quantitative measurements like the production of coal, iron, textiles, and steam power and on national income, but the proud nation also trailed England and Germany in newer areas like steel, electricity, and chemicals. These numbers have prompted historians to search for explanations. Some point to the economic disadvantages of scarce coal deposits, sluggish population growth, and agricultural inefficiency. The inhibiting effect of high tariffs on technological innovation also receives emphasis. Others blame the political instability of nineteenth-century France and the debilitating effect of its loss to Germany in 1870—for the victors seized Alsace, with its thriving textile industries, and Lorraine, with its rich deposits of iron ore, then imposed a stiff war indemnity on the loser. Yet another school argues stridently that French entrepreneurs were typically more interested in social elevation to the upper class than in plowing back profits into their businesses (Clapham 1966; Dunham 1955; Kindleberger 1964; Landes 1969).

Although most of these arguments have some merit, a different and more positive image of France appears if we take a closer look at technology. Such an examination reveals a nation of highly sophisticated scientists and engineers whose ingenuity created "niches" in novelty areas, luxury markets, or where France enjoyed comparative advantages. Thus she was the largest European producer of automobiles and carburetors and the world leader in aircraft frames and engines in 1913. Its scientists—notably H. E. Sainte-Claire Deville (1870s) and Paul L. T. Hèroult (1880s)—pioneered the electrolytic process for producing metallic aluminum. Blessed with some of the richest deposits of high-grade bauxite in Europe, France produced nearly twice as much aluminum as England in 1913—and more than sixteen times the German total. French steel output amounted to only 31 percent of the German but mirrored its great rival technologically, with roughly two-thirds produced by the "basic" Bessemer method, the remainder with "basic" Siemens-Martin. Although first patented in Germany, moreover, the electric-arc steel furnace was improved by Hèroult in the 1890s, and by the eve of World War I, 50 percent of the world's electric-arc furnaces were manufactured in France. Finally, while only one of France's eighty-seven sulfuric acid works used the contact process in 1913, her lead-chamber plants turned out nine hundred thousand tons. This nearly equaled Britain's annual output that year—but since 1870 France's production of sulfuric acid had grown four times as fast (Chadwick 1958; Gille 1986, vol. 1; Haber 1971; Multhauf 1967).

The French artificial-soda industry of the late 1800s developed in similarly dynamic fashion. The process that dominated the First Industrial Revolution was the work of the Frenchman Nicolas Leblanc in the early 1790s (along with English industrialists who made the procedure commercially feasible after the Napoleonic Wars). Common salt and sulfuric acid reacted on a lead tray in the first stage of the operation to produce sodium sulfate. Further heating with chalk yielded the desired product of sodium carbonate. By mid-century immense capital had been sunk in complex chemical works utilizing this process, especially in England, for soda was an essential ingredient of soap, glass, and saltpeter. In the 1860s, however, the Belgian Ernest Solvay discovered a superior method. Ammoniacal salt brine mixed with carbon dioxide gas in an elaborate reaction tower under controlled pressure and temperature to produce sodium bicarbonate, which further heating reduced to soda. Because it was a continuous process that used materials more efficiently, Solvay soda undercut Leblanc in price by 20 percent in the 1870s. The French converted to the new process quicker than any other nation, becoming the fastest-growing soda industry in Europe during the prewar decades. Behind the massive Solvay and Company, which accounted for two-thirds of national output, the French soda industry pulled close to Germany, with about four hundred thousand tons by 1913, trailing only England, with seven hundred fifty thousand tons (Haber 1971; Landes 1969; Multhauf 1967).

It is this combination of relative economic decline in the aggregate—generally failing to catch up with England as well as dropping behind Germany—with a technological and scientific élan vital that propelled her to the world's forefront in certain categories that makes France something of a conundrum for historians of the late nineteenth century. Some credible answers to the riddle have been sketched at the outset of this passage: coal was scarce; political culture was not conducive to long-term investments; and Germany crushed France in 1870, thereby interrupting growth for years. Reflecting a cultural preference for smaller families, moreover, French population grew a scant 44 percent over the long century from 1800 to 1913—far behind England's demographic expansion of 508 percent and Germany's population growth of 276 percent over the same period. Compounding the depression of demand that this induced was a supply situation in the French countryside. Because farm plots remained small and unenclosed until the 1890s and beyond, agricultural efficiency improved very slowly over the century, and workers were required in the countryside. Thus 42 percent of the labor force still worked on farms in 1913, compared with 34 percent in Germany and only 8 percent in England. These were the economic constraints that French scientific and engineering minds had to combat. And they contended successfully, for in absolute terms France still managed to grow and prosper. Only in certain comparative categories, in other words, was France slipping. Nevertheless, in an age of violence, arms races, and a seemingly inevitable world war, such international comparisons raised serious questions about national security.

THE RELATIVE DECLINE OF ENGLAND

The economic "decline" of Victorian England has always been a controversial topic. It moved center stage politically in 1903 when Joseph Chamberlain, the colonial secretary, initiated his campaign for a system of protective tariffs that favored the nations of the British Commonwealth. The world had changed with the rise of Germany and the United States, he asserted, and Britain could no longer afford free-trade policies that left her industries vulnerable to the challengers. Chamberlain's movement triggered an angry response from indignant patriots who charged that it was not true. English entrepreneurs who had conquered the industrial world a century ago needed no artificial props to stay on top. It was a classic struggle of conflicting economic interests as import-export businesses opposed measures that threatened the free flow of trade, while industrialists more dependent on the home market welcomed a proposal that undercut the foreign competition. By 1910, however, Chamberlain conceded political defeat. The secretary's call to arms had offended Englishmen whose wounded pride would not let them admit that he had been right.

To be sure, there were still many examples of economic and technological vitality in the world's workshop on the eve of the Great War. London was the financial and insurance capital of the world, and the pound sterling was the world's linchpin currency. England still mined more coal and produced more soda than Germany, and its textile output and merchant marine were four and five times as great, respectively, as Germany's. English steam engine horsepower continued to exceed that of the German Empire by almost two to one, moreover, and electrical capacity, when measured on a per capita basis, was also larger. Indeed Englishman Charles Parsons invented a new type of steam engine in 1884 to drive electrical generators. The so-called steam turbine avoided the accumulating physical problems of alternating mechanical motion by moving directly to rotary motion. Jets of steam drove against a series of vanes or buckets that branched off in stages from a rotating axis. The turbine attained efficiencies of 70–80 percent, required half the space, and

Lands of Unlimited Possibilities   103

Figure 3.8: Size comparison of generator with reciprocating steam engine (*bottom*) and generator with steam turbine. From Conrad Matschoss, *50 Jahre Berliner Elektrizitätswerke, 1884–1934* (n.d.), p. 132.

by 1907 was driving maritime engines of sixty-eight thousand horsepower. Parsons also developed an electrical generator that ran at eighteen thousand revolutions per minute (see figure 3.8). Together his turbine and generator "represent the greatest innovation in the use of steam power since Watt's construction of an engine to produce rotary power," writes David Landes. "They also made possible an efficient, large-scale electrical power industry" (1969, 279).

These positive indicators mixed incongruously and uneasily, however, with disturbing and unmistakable signs of economic and technological decline. The story of the British soda industry is a good example of these mixed signals. Having invested fortunes over two generations in the Leblanc manufacturing process, entrepreneurs believed that it was reasonable to attempt to compete with the Solvay method. By reducing costs, installing more efficient machinery, earning extra profits from marketing by-products like chlorine, and merging smaller companies into the huge United Alkali Company, Britain's Leblanc producers remained profitable until the 1890s. During that decade, however, high tariffs abroad plus more efficient electrolytic processes for making chlorine undercut Leblanc's profit margins. But the big company refused to adopt the Solvay method. In stark contrast to the conservatism of United Alkali was the dynamic firm of Brunner and Mond, which shot to the top of the island's soda industry by choosing the newer technology. Nevertheless, England still produced about 6 percent of its soda with the antiquated Leblanc method in 1913—the highest percentage among the world's industrial nations (Haber 1971; Landes 1969).

The steel industry is another representative example. England was in the forefront of nations attempting to replace iron as the primary material of the technological system. In what proved an initial advantage, she possessed deposits of nonphosphoric hematite ore that facilitated adoption of the original Bessemer process. Then, when the Siemens-Martin open-hearth procedure spread in the 1870s, England was the quickest to adopt it. The slowness of the process enabled furnace masters to decarburize iron with the precision required of steel plate destined for England's huge shipbuilding sector. Over 79 percent of Britain's seven million tons of steel in 1913 was Siemens-Martin. What strikes historians of technology, however, is the small scale and relative inefficiency of the British plant. German open-hearth furnaces and Bessemer converters were much larger—the former five or six times the size of Britain's. Contributing to this disadvantage was England's adherence to "acid" steel. For the Gilchrist-Thomas process that swept the continent, allowing steelmakers to use phosphoric ores, did not impress British entrepreneurs who had access to hematite ores. The siliceous (i.e., acidic) bricks of English furnaces quickly broke down when contacted by basic ores, however, thus necessitating costly and time-consuming replacement, which tended to limit scale. Residue from the deteriorating bricks also allowed impurities to enter the metal. England's steel industry was behind, in other words, both quantitatively and qualitatively (Landes 1969).

A glance at the aggregate statistics brings England's otherwise contradictory mixture of economic and technological indicators into sharper focus. The British industrial sector grew at an annual average rate of 1.9 percent from 1870 to 1913. By comparison, France rose at 3.1 percent and Germany at 4.9 percent over the same decades. Even more damning is the figure for the increase in gross national product per man hour—the standard measure of technological change. While Germany grew at 2.1 percent and France at 1.8 percent, England expanded more slowly at 1.5 percent. An isolated examination of the years after 1900, moreover, reveals a shocking absolute decline in productivity per man hour. What was happening to this great industrial nation (Kindleberger 1964)?

Paradoxically, a variety of circumstances induced many British entrepreneurs to follow a short-run economic rationality, which undermined technological change and competitiveness in the long run. In the case of British adherence to the Leblanc soda procedure, for instance, gargantuan investments and the related fixed or "sunk" costs of one technology made it imprudent—or economically "irrational"—to scrap this plant, especially if profit margins remained positive. After thirty years the older method succumbed, but in the meantime potential productivity gains were lost. Similarly, British steel producers refused to abandon their acidic technology while it was still profitable, despite the accumulating evidence that Gilchrist-Thomas made better steel. There are other examples of this interesting phenomenon of short-run rationality and long-run short-sightedness. Thus many English factories were situated in crowded urban neighborhoods that did not permit larger operations. When faced with the option of reinvesting in new plant and machinery in a suburban or rural site, many businessmen decided to forego the risks—and potential earnings from economies of scale—as long as the existing plant was in the black.

England also experienced institutional and political problems that slowed its transition from the old technological system to the new. A relatively fragmented pattern of ownership inhibited adoption of new techniques, for example, when two or more owners had to agree on the change. Thus dock owners wanted to install electrical cranes but were prevented by shipowners, who refused to make onboard changes to accommodate the device. In Germany the tendency was toward "vertically integrated" firms that owned operations from one end of the technical concatenation to the other. With the most powerful trade union movement in the world, Great Britain also suffered when organized labor resisted technological displacement of workers. The examples are numerous: tailoresses opposed button-holing machines, bottle-makers fought bottling machines, dockworkers electrical cranes, and so on. Or machinists fought layoffs when new turret lathes operated by one man could do the work of many. From 1900 to 1914, in fact, 13 percent of all striking workers struck for gains of this sort—21 percent in the engineering, metal, and shipbuilding industries. Political difficulties arose after 1906, moreover, when the Liberal Party introduced social reforms financed by taxes that wealthy Englishmen perceived as "soak the rich" measures. As a result, capital flowed out of the country, annually exceeding domestic investment by twelve million pounds from 1906 to 1912. Thus some entrepreneurs may have been cut off from funds required for technological modernization. But if this outflow—as appears equally likely—was not cause, but rather effect, then it also reflected entrepreneurial indifference to new investments. However one interprets this phenomenon, the implications were negative (Frankel 1955; Landes 1969; Levine 1967; McCloskey and Sandberg 1971).

Finally, there is reason to conclude that England, the land that had taught others to appreciate science, failed to maintain an adequate flow of scientific knowledge to industry. We must not forget the very dynamic nature of the nineteenth-century economy. It was a rapidly unfolding age of electricity, chemistry, motors, and ingenious, ever more complex and *scientific* technologies (see below). After mid-century, British investigators began to express respect for European systems of education that seemed better equipped to serve this new type of industry. One team was startled to discover in 1872 that there were more highly trained chemists at the University of Munich alone than in all of Great Britain. Another royal commission made the following observation in 1884:

> When, less than a half century ago, continental countries began to construct railways and to erect modern mills and mechanical workshops, they found themselves face to face with a full-grown industrial organization in this country. . . . To meet this state of things, foreign countries established technical schools . . . [which] now exist in nearly every continental state, and are the recognized channel for the instruction of those who are intended to become the technical directors of industrial establishments. Many of the technical chemists have, however, been, and are being, trained in the German universities. Your commissioners believe that the success which has attended the foundation of extensive manufacturing establishments, engineering shops, and other works, on the continent, could not have been achieved to its full extent in the face of retarding influences, had it not been for the system of high technical instruction in these schools, for the facilities for carrying on original scientific investigation, and for the general appreciation of the value of that instruction, and of original research, which is felt in those countries (Ashby 1958, 786–87).

This indirect criticism of classical education in English public schools and universities fell on deaf ears, however, for highly placed aristocrats and intellectuals, appalled by the ugly social consequences of industrialization, successfully resisted any curricular changes that would aid industry, the culprit. "This development set England apart from its rivals," writes Martin Wiener, "for in neither the United States nor Germany did the educational system encourage a comparable retreat from business and industry" (1981, 24). Additional evidence emanating from the laboratories of Germany's universities and giant industrial firms in the 1880s and 1890s finally galvanized a parliamentary majority to take action in 1902. The establishment that year of a state system of secondary education with an emphasis on technical instruction was the best testimony, perhaps, to a century of educational negligence. England was clearly struggling now to salvage something from its once seemingly insurmountable lead.

THE UNITED STATES' CONTRIBUTION TO THE SECOND INDUSTRIAL REVOLUTION

Like France and Germany at mid-century, the United States of America could not match the industrial might of England. Production was a small fraction of the world's leader in every category except one: US steam engine capacity was 30 percent higher, reflecting the rapid expansion of steam power on rivers and railroads. By 1913, however, the United States had emerged as an economic and technological superpower, far surpassing Europe's individual leaders in almost every way. Steel output in the United States nearly equaled (94 percent) the total for Germany, England, France, Russia, and Italy. Aluminum production topped England, France, and Germany by 12 percent. Output of sulfuric acid came close (89 percent) to that of Germany and England combined, while electrical energy surged 73 percent above the sum of all Europe. US petroleum production exceeded that in Europe—primarily Russia—by three-and-a-half times, and its five hundred thousand automobiles more than quintupled France, England, and Germany together (96,300). But the United States was more than a quantitative success—its economy also ran very efficiently. Thus overall output per man hour rose 2.4 percent annually from 1870 to 1913, above Germany (2.1 percent), France (1.8 percent), and England (1.5 percent).

As we know, greater efficiency usually reflects more technology. While this was certainly true in the American case, it is interesting that most of the ideas—with notable exceptions like Bell's telephone, Edison's lightbulb, Maxim's machine gun, and the Wrights' airplane—came from Europe. The great breakthroughs in metallurgy, steam power, chemistry, and electricity originated in England, France, and Germany, not in the land of Yankee ingenuity. There was another American exception, however, that went beyond novelty and invention to make a unique contribution to the new technological order. Indeed, Europeans had observed as early as the 1850s that Americans were evolving a special approach to industrial production driven by mass markets and based on precision machine-tooled, interchangeable parts. It is now time to discuss this "American system" of manufacturing.

Our story begins at the government armory in Springfield, Massachusetts, in the early 1800s. The US Army wanted to produce firing mechanisms, barrels, and gun stocks of a standard sort to replace the hand-filed and custom-fitted weapon parts—each a little different from the next—that were almost impossible to replace on the battlefield. The armorers faced none of the normal businessman's anxiety over short investment capital or limited sales of items produced, for funding and the market were guaranteed by the state. As the decades unfolded to the 1860s, workers achieved such high levels of precision that gun parts could be interchanged.

Two innovations from the armory made this possible: the universal milling machine and the turret lathe. The former replaced hand-chiseling and filing of parts and dies requiring intricate curves, cross sections, and spirals. Employing rotary cutters, the milling machine worked rapidly, efficiently, and accurately. The turret lathe was used mainly for cutting screws and other metal parts. It held a cluster of tools on a vertical or horizontal axis that rotated, making it possible to perform a sequence of cuts on a piece without removing the workpiece or retooling the lathe. Like the universal milling machine, this lathe combined the advantages of speed and precision. What is more significant, the armory opened its doors to curious craftsmen and factory owners who were soon employing the same novel techniques. After the Civil War, interchangeable parts spread to watch and clock making as well as the manufacture of sewing machines, cash registers, typewriters, farm tools, bicycles, locomotives, and automobiles.

In the process, a third revolutionary machine tool came to the American system: the precision grinding machine, accurate to a fraction of a thousandth of an inch. First employed on sewing machines in the 1850s and 1860s, grinding was especially useful in the nascent automobile industry after 1900, for in a few minutes heavy grinders removed unwanted excess material from crank and cam shafts, ball bearings, piston rings, and cylinders. It was along the assembly lines of Michigan, of course, where the system of interchangeable parts came famously into its own (see figure 3.9). But all these industries followed the same economic and technological logic: mass markets called for mass production, which, in turn, required the replacement of hand techniques by mechanization and standardization. Universal milling machines, turret lathes, and grinders were the breakthrough techniques that made the whole system feasible (Hounshell 1984; Rosenberg 1972; Merritt Roe Smith 1977; Woodbury 1967).

By the late 1800s some of the most ingenious technology from America was beginning to flow back into Europe. In steel production, for instance, the immense scale of the US market led to the development of huge one-hundred- to three-hundred-ton open hearth furnaces that were capable of tilting to pour out the finished metal. German steelmakers in the Ruhr followed suit. German armaments firms like Loewe and Company also emulated the technique of interchangeable parts. Able to borrow selectively like this, the empire could face economic developments across the Atlantic calmly and confidently—as long as the real scientific momentum behind the Second Industrial Revolution remained German.

In 1903, however, Germans paid serious attention to a statement by Henry Pritchett, president of the Massachusetts Institute of Tech-

Lands of Unlimited Possibilities   111

Figure 3.9: Part of the assembly-line process at Highland Park, Michigan. Courtesy of Ford Motor Company.

nology: "Germans need fear in the industrial world neither the Englishman nor the Frenchman, only the American." What made the boast especially worrisome to German industrialists was the rapid ascension of the United States as the world's leader in research and development. Thus America's top twelve universities expended double the R&D budgets of Germany's twenty-one universities—and the rate of growth was also twice as fast. In 1901–02, moreover, private capital founded the Rockefeller Institute for Medical Research and the Carnegie Institution. Each was capitalized at about ten million dollars, and together they generated annual interest equal to the budget of a German university. The government added momentum to the US scientific challenge with the establishment of the National Bureau of Standards in 1901. Among its many functions was a chemistry section budgeted at twice the level of a German university. Now we can begin to understand the urgency in Emil Fischer's voice as he addressed his peers and his sovereign in 1911. Without more scientific research, qualitative leadership of the industrial world would change hands again.

THE NEW TECHNOLOGICAL SYSTEM AND THE COMING OF WORLD WAR I

The history of science's contribution to the evolving technological system of the late nineteenth century underscored the logic behind Fischer's fears. His own field, chemistry, was a prime example. Since the First Industrial Revolution, researchers had discovered many of the elements that compose today's periodic table. Scientists also probed the molecular structure of the elements and more complex organic substances to determine the number and kind of atoms within them. Once these structures were known, it was only a short (albeit painstaking) step to the creation of synthetic substances like dyestuffs.

Electricity, another technology central to the system, was also a series of purely scientific breakthroughs. Thus electrical batteries stemmed from the discovery by Alessandro Volta in 1800 that certain metals like silver and zinc generated a current when brought together in a weak acid solution. And generators were based on scientific experiments by Christian Oersted in 1820 and Michael Faraday in the 1840s that demonstrated, first, that a magnetic field exists around an electric conductor and, second, that conductors and magnets created electrical current when moved in relation to one another. Decades of follow-up work was required—as the earlier quote of Professor Landes demonstrates—before myriad applications were found for the new source of power and energy. The important point that bears repeating here is that generations of scientific, curiosity-driven research lay behind the emergence of the electrical industry (Dibner 1967).

Metallurgy also benefited from science during the late 1800s. The previous century's work, as already noted, had led to the discovery of most of the elements. Once carbon, sulfur, manganese, silicon, and phosphorous were known, their properties—and often deleterious effects in iron and steel—could be studied systematically. Metallurgists also conducted experiments on the microcrystalline structure of iron and steel that revealed the reformative impact of hammering and rolling on crystal patterns, the recrystallizing effects of heating, and the exact temperatures when these changes occurred. After the introduction of x-ray diffraction in 1912, researchers could determine the exact arrangement of atoms within ferrous crystals. Finally, scientists developed a mathematical theory of elasticity after thousands of tests on the resistance of iron and steel to bending and stretching. In contrast to chemical and electrical developments, where technology drew on existing bodies of scientific knowledge, metallurgical science largely followed technology, attempting to explain chemical reactions and physical changes. Nevertheless, this empirical approach led to tremendous improvements in metallurgical practice (Smith 1967a).

We can further demonstrate the importance of science to late-nineteenth-century technology by examining the centrality and integration of the newer developments—their convergence and concomitance—within the technological system. Powered by batteries, steam turbines, waterfalls, and dams, for example, electrical dynamos and generators transmitted energy over ever-widening networks to a complex web of technical structures and ensembles. Electricity drove streetcars and started automobile and airplane engines. It lit homes, workshops, offices, factories, and streets and made possible the telegraph, telephones, wireless telegraphy, and the first vacuum-tube radio. Electrolytic processes produced caustic soda, chlorine, and aluminum, and electrical current made the purest steel. In tandem with steam turbines and the older reciprocating engines, electricity solved society's power requirements, facilitating one of the most dynamic expansions in history. Indeed between 1870 and 1913 world energy use (from all sources) skyrocketed from 1.67 billion to 10.84 billion megawatt hours. Over one-fifth of the prime mover capacity produced from this energy in 1913—21.8 percent in England, 22.7 percent in Germany—came from electrical generators (Gille 1986, 1:683; Landes 1969).

In turning to chemicals, we find an old technical ensemble, the lead-chamber process for producing sulfuric acid, and a new one—the contact method for making both sulfuric and nitric acid—combining to meet society's needs. In addition to surviving traditional uses like gunpowder, matches, paper, and soap, these acids now helped to refine petroleum, convert nitrates into nitric acid, make nitrate fertilizers that replaced expensive organic products (e.g., guano), and produce the powerful nitrogen-based high explosives. The chemical industry also created a host of synthetic materials that were applied in numerous technical concatenations within the system: dyes in textiles, dynamite in construction, and aspirin and medicinal products in healthcare.

The metallurgical concatenations provide additional examples of technical convergence and concomitance. The discussion of American machine tools needs little elaboration here. Crucial complex technical structures like the universal milling machine, the turret lathe, and the grinding machine facilitated a system of interchangeable metal parts that spread from gun making to sewing machines, bicycles, and automobiles. And the metal of choice by 1900 (for machine tools as well as the products they fashioned) was steel—and steel alloys—produced with laboratory-regulated consistency. In fact, while iron continued to find uses in home and industry, steel spread throughout the economies of the developed nations. The cheap, strong, flexible material replaced tens of thousands of miles of rails in Europe and the United States. It provided the metal for electrical cables; for reinforced bridges, towers, and concrete skyscrapers; as well as for the superheated, high-pressure "bomb" that contained the contact acid process.

It also came to be a barometer of war-making capacity—shells, rifle barrels, artillery pieces, armor plating, and the huge 125-ton turrets on the new "Dreadnought" battleships were all made of steel. Indeed we gain a deeper appreciation for the technical convergences of the era—and their ominous historical significance—if we recall that the machine-tooled shells of 1914 were packed with the new chemical explosives, loaded into sophisticated naval guns fired electrically, onboard ships driven to their deadly destinations by steam turbine engines.

The earlier technological system "mutated" during the late nineteenth century into a new series of interlocking technologies. As scientists and engineers removed one technical bottleneck after another, the tentacles of the new system spread across ensembles and concatenations—as well as across international borders. The hostile nations of Europe eyed one another warily as the decades drew on to 1914, anxiously watching for any sign that enemies had technological advantages. A great arms race ensued as soldiers in every nation discarded older notions of bravado and accustomed themselves to the idea of fighting with machines. Rapid-firing artillery and magazine rifles, machine guns, the first airplanes, and the great new battleships appeared in European arsenals. Germany—whose industries grew at the fastest pace, whose scientists and engineers generated many of the new technologies, and whose leaders were blunt and arrogant—alarmed England and France, two proud powers declining relative to Germany. Russia, struggling desperately to enter the new technological age, allied itself with England and France. "The most determined efforts of the wisest men," concludes David Landes, "did not avail to appease the resentments and enmities that grew out of the consequently altered balance of power" (1969, 248). Those students of history who see the resultant war as "the thrashing of a [capitalist economic] system in the process of decline and dissolution" are wrong, he continues. "The fact is that these were the growing pains of a [technological] system in process of germination" (Gille 1986, 1:693).

The result was a buildup of international tension that seemed to keep pace with the surging, six-and-a-half-fold increase in energy output (see above). One young man, the novelist Stefan Zweig, explained the outbreak of war in just this way. The explosion "had nothing to do with ideas and hardly even with frontiers," he recalled. "I cannot explain it otherwise than by this surplus force, a tragic consequence of the internal dynamism that had accumulated in forty years of peace and now sought violent release" (Tuchman 1966, xv).

EPILOGUE

Four years—and thirty-seven million casualties—later, it was finally over. Fittingly enough, the growth of the technological system that contributed to the Great War's coming accelerated from 1914 to 1918. The volume and velocity of messages greatly intensified as generals sent orders down the chain of command. As a result, telephone technology matured and the age of radio began. Flimsy biplanes made of wood, canvas, and wire in 1914 began the metamorphosis to the fast aluminum monoplanes of the postwar period. Germany made the production of nitric acid from atmospheric nitrogen a practical engineering reality, thus undermining older production methods dependent on nitrate ores. And output of electricity tripled, transmitted within quickly spreading regional distribution grids over lines operating at 110,000 volts. Consequently, steam engines receded increasingly in importance, representing only about a third of prime mover capacity by 1929 (Hughes 1983; Landes 1969).

The technological system—Gille dubs it the "modern" system—reached the limits of its potential during the second great world conflagration. From 1939 to 1945 the older system began to mutate into our "contemporary" system as war gave birth to jets, rockets, computers, and atomic power. It is a story that is still unfolding, however, and therefore its telling is better left to others.

REFERENCES

Alder, Ken. 1997. *Engineering the Revolution: Arms and Enlightenment in France, 1763–1815*. Princeton, NJ: Princeton University Press.

Allen, J. S. 1977. *The Steam Engine of Thomas Newcomen*. New York: Science History Publications.

Armytage, W. H. G. 1965. *The Rise of the Technocrats: A Social History*. London: Routledge and Kegan Paul.

Ashby, Eric. 1958. "Education for an Age of Technology." In *A History of Technology*, edited by Charles Singer, E. J. Holmyard, A. R. Hall, and T. I. Williams. Vol. 5: *The Late Nineteenth Century 1850 to 1900*. Oxford: Clarendon Press.

Ashton, T. S. 1955. *An Economic History of England: The 18th Century*. London: Methuen.

Bacon, Francis. 1955a. *The Advancement of Learning*. In *Selected Writings of Francis Bacon*, edited by Hugh G. Dick. New York: Random House.

———. 1955b. *The New Atlantis*. In *Selected Writings of Francis Bacon*, edited by Hugh G. Dick. New York: Random House.

Beer, John J. 1959. *The Emergence of the German Dye Industry*. Urbana: University of Illinois Press.

Berg, Maxine. 1994. *The Age of Manufacturers 1700–1820*. London: Routledge.

Bonney, Richard. 1991. *The European Dynastic States 1494–1660*. Oxford: Oxford University Press.

Brose, Eric Dorn. 1985. "Competitiveness and Obsolescence in the German Charcoal Iron Industry." *Technology and Culture* 26, no. 3 (July): 532–59.

———. 1997. *German History 1789–1871: From Holy Roman Empire to Bismarckian Reich*. Providence, RI: Berghahn Books.

———. 1993. *The Politics of Technological Change in Prussia: Out of the Shadow of Antiquity, 1809–1848*. Princeton, NJ: Princeton University Press.

Bruchey, Stuart. 1968. *The Roots of American Economic Growth 1607–1861*. New York: Harper and Row.

Butterfield, Herbert. 1957. *The Origins of Modern Science 1300–1800*. New York: Free Press.

Chadwick, R. 1966. "New Extraction Processes for Metals." In *A History of Technology*, edited by Charles Singer, E. J. Holmyard, A. R. Hall, and T. I. Williams. Vol. 5: *The Late Nineteenth Century 1850 to 1900*. Oxford: Clarendon Press.

Chambers, J. D. 1953. "Enclosure and Labour Supply in the Industrial Revolution." *Economic History Review* 5, no. 3: 319–43.

Clapham, J. H. 1966. *Economic Development of France and Germany 1815–1914*. Cambridge: Cambridge University Press.

Clow, A., and N. L. Clow. 1958. "The Chemical Industry: Interaction with the Industrial Revolution." In *A History of Technology*, edited by Charles Singer, E. J. Holmyard, A. R. Hall, and T. I. Williams. Vol. 4: *The Industrial Revolution 1750 to 1850*. Oxford: Oxford University Press.

Crafts, N. F. R. 1985. *British Economic Growth during the Industrial Revolution*. Oxford: Oxford University Press.

Crafts, N. F. R., and C. K. Harley. 1992. "Output Growth and the British Industrial Revolution: A Restatement of the Crafts-Harley View." *Economic History Review* 45: 703–30.

Crouzet, François. 1967. "England and France in the Eighteenth Century: A Comparative Analysis of Two Economic Growths." In *The Causes of the Industrial Revolution in England*, edited by R. M. Hartwell. London: Meuthen.

Cuenca Esteban, Javier. 1997. "The Rising Share of British Industrial Exports in Industrial Output." *Journal of Economic History* 57: 879–906.

Davis, Ralph. 1979. *The Industrial Revolution and British Overseas Trade.* Leicester: Leicester University Press.

Deane, Phyllis. 1967. "The Industrial Revolution and Economic Growth: The Evidence of Early British National Income Estimates." In *The Causes of the Industrial Revolution in England,* edited by R. M. Hartwell. London: Meuthen.

Deane, Phyllis, and W. A. Cole. 1969. *British Economic Growth 1688–1959.* Cambridge: Cambridge University Press.

De Vries, Jan. 1993. "Between Purchasing Power and the World of Goods: Understanding the Household Economy in Early Modern Europe." In *Consumption and the World of Goods,* edited by J. Brewer and R. Porter. London: Routledge.

———. 1994. "The Industrial Revolution and the Industrious Revolution." *Journal of Economic History* 54: 249–70.

Dibner, Bern. 1967. "The Beginning of Electricity." In *Technology in Western Civilization,* edited by Melvin Kranzberg and Carroll W. Pursell Jr. Vol. 1: *The Emergence of Modern Industrial Society Earliest Times to 1900.* New York: Oxford University Press.

Dickinson, H. W. 1958. "The Steam Engine to 1830." In *A History of Technology,* edited by Charles Singer, E. J. Holmyard, A. R. Hall, and T. I. Williams. Vol. 4: *The Industrial Revolution 1750 to 1850.* Oxford: Oxford University Press.

Dickson, P. G. M. 1967. *The Financial Revolution in England.* New York: St. Martin's Press.

Dunham, Arthur Louis. 1955. *The Industrial Revolution in France 1815–1848.* New York: Exposition Press.

Ellis, Aytoun. 1956. *The Penny Universities: A History of the Coffee Houses.* London: Secker and Warburg.

Engerman, Stanley. 1994. "Mercantilism and Overseas Trade, 1700–1800." In *The Economic History of Britain Since 1700,* edited by R. Floud and D. McCloskey. Vol. 1. Cambridge: Cambridge University Press.

Ferguson, Eugene S. 1967a. "Metallurgical and Machine-Tool Developments." In *Technology in Western Civilization,* edited by Melvin Kranzberg and Carroll W. Pursell Jr. Vol. 1: *The Emergence of Modern Industrial Society Earliest Times to 1900.* New York: Oxford University Press.

———. 1967b. "The Steam Engine before 1830." In *Technology in Western Civilization,* edited by Melvin Kranzberg and Carroll W. Pursell Jr. Vol. 1: *The Emergence of Modern Industrial Society Earliest Times to 1900.* New York: Oxford University Press.

———. 1967c. "Steam Transportation." In *Technology in Western Civilization*, edited by Melvin Kranzberg and Carroll W. Pursell Jr. Vol. 1: *The Emergence of Modern Industrial Society Earliest Times to 1900*. New York: Oxford University Press.

Firth, Katherine R. 1979. *The Apocalyptic Tradition in Reformation in Britain, 1530–1645*. Oxford: Oxford University Press.

Forbes, R. J., and Cyril Stanley Smith. 1957. "Metallurgy and Assaying." In *A History of Technology*, edited by Charles Singer, E. J. Holmyard, A. R. Hall, and T. I. Williams. Vol. 3: *From the Renaissance to the Industrial Revolution*. Oxford: Clarendon Press.

Forbes, R. J. 1957. "Power to 1850." In *A History of Technology*, edited by Charles Singer, E. J. Holmyard, A. R. Hall, and T. I. Williams. Oxford: Oxford University Press.

Frankel, Marvin. 1955. "Obsolescence and Technological Change in a Maturing Economy." *American Economic Review* 45: 256–315.

Fremdling, Rainer. 1975. *Eisenbahnen und deutsches Wirtschaftswachstum 1840–1879: Ein Beitrag zur Entwicklungstheorie und zur Theorie der Infrastruktur*. Dortmund: Gesellschaft für Westfälische Wirtschaftsgeschichte e.v.

Gagliardo, John. 1991. *Germany under the Old Regime*. London: Longman.

Gerschenkron, Alexander. 1994. "Economic Backwardness in Historical Perspective." In *The Industrial Revolution in Europe*, edited by P. K. O'Brien. Vol. 1. Oxford: Blackwell.

Gilbert, K. R. 1958. "Machine Tools." In *A History of Technology*, edited by Charles Singer, E. J. Holmyard, A. R. Hall, and T. I. Williams. Vol. 4: *The Industrial Revolution 1750 to 1850*. Oxford: Oxford University Press.

Gille, Bertrand. 1966. *Engineers of the Renaissance*. Translated by J. Brainch et al. Cambridge, MA: MIT Press.

———. 1986. *The History of Techniques*. Translated by P. Southgate and T. Williamson. 2 vols. New York: Gordon and Breach.

Gribbin, John. 2002. *Science: A History 1543–2001*. London: Allen Lane.

Haber, L. F. 1971. *The Chemical Industry 1900–1930: International Growth and Technological Change*. Oxford: Clarendon Press.

Hall, A. R. 1957. "Military Technology." In *A History of Technology*, edited by Charles Singer, E. J. Holmyard, A. R. Hall, and T. I. Williams. Vol. 3: *From the Renaissance to the Industrial Revolution*. Oxford: Oxford University Press.

Hall, A. Rupert. 1967. "Early Modern Technology, to 1600." In *Technology in Western Civilization*, edited by Melvin Kranzberg and Carroll W. Pursell Jr. Vol. 1: *The Emergence of Modern Industrial Society Earliest Times to 1900*. New York: Oxford University Press.

Hall, Manly P. 1937. *Freemasonry of the Ancient Egyptians*. Los Angeles: Philosophical Research Society.

Hans, Nicholas Adolph. 1951. *New Trends in Education in the Eighteenth Century*. London: Routledge and Kegan Paul.

Harley, C. K. 1982. "British Industrialization before 1841: Evidence of Slower Growth during the Industrial Revolution." *Journal of Economic History* 42: 267–89.

Holborn, Hajo. 1964. *A History of Modern Germany, 1648–1840*. New York: Alfred A. Knopf.

Holmyard, E. J. 1966. "Dyestuffs in the Nineteenth Century." In *A History of Technology*, edited by Charles Singer, E. J. Holmyard, A. R. Hall, and T. I. Williams. Vol. 5: *The Late Nineteenth Century 1850 to 1900*. Oxford: Oxford University Press.

Hounshell, David A. 1984. *From the American System to Mass Production: The Development of Manufacturing Technology in the United States*. Baltimore: Johns Hopkins University Press.

Hudson, Pat. 1992. *The Industrial Revolution*. London: E. Arnold.

———. 1996. "Proto-Industrialization in England." In *European Proto-Industrialization*, edited by Sheilagh C. Ogilvie and Markus Cerman. Cambridge: Cambridge University Press.

Hughes, Thomas P. 1983. *Networks of Power: Electrification in Western Society, 1880–1930*. Baltimore: Johns Hopkins University Press.

Hyde, Charles K. 1977. *Technological Change and the British Iron Industry 1700–1870*. Princeton, NJ: Princeton University Press.

Jacob, Margaret C. 1981. *The Radical Enlightenment: Pantheists, Freemasons and Republicans*. London: Allen and Unwin.

———. 1988. *The Cultural Meaning of the Scientific Revolution*. New York: Alfred A. Knopf.

———. 1997. *Scientific Culture and the Making of the Industrial West*. New York: Oxford University Press.

Jarvis, C. Mackechnie. 1966a. "The Distribution and Utilization of Electricity." In *A History of Technology*, edited by Charles Singer, E. J. Holmyard, A. R. Hall, and T. I. Williams. Vol. 5: *The Late Nineteenth Century 1850 to 1900*. Oxford: Oxford Clarendon Press.

———. 1966b. "The Generation of Electricity." In *A History of Technology*, edited by Charles Singer, E. J. Holmyard, A. R. Hall, and T. I. Williams. Vol. 5: *The Late Nineteenth Century 1850 to 1900*. Oxford: Oxford University Press.

Jensen, Merrill. 1950. *The New Nation: A History of the United States during the Confederation 1781–1789*. New York: Random House.

Joest, Hans-Josef. 1982. *Pioneer im Ruhrgebiet*. Stuttgart: Seewald Verlag.

Johnson, Jeffrey Allan. 1990. *The Kaiser's Chemists: Science and Modernization in Imperial Germany*. Chapel Hill: University of North Carolina Press.

Jones, Richard Foster. 1965. *Ancients and Moderns: A Study of the Rise of the Scientific Movement in Seventeenth Century England*. Berkeley: University of California Press.

Kearney, Hugh. 1986. *Science and Change 1500–1700*. New York: McGraw-Hill.

Keller, Ludwig. 1918. *Die Freimaurerei: Eine Einführung in ihre Anschauungswelt und ihre Geschichte*. Leipzig: B. G. Teubner.

Kindleberger, Charles P. 1964. *Economic Growth in France and Great Britain, 1851–1950*. Cambridge, MA: Harvard University Press.

König, Wolfgang. 1990. "Massenproduktion und Technikkonsum: Entwicklungslinien und Triebkräfte der Technik zwischen 1880 und 1914." In *Propyläen Technikgeschichte*, edited by Wolfgang Konig. Bd. 4: Netzwerke Stahl und Strom 1840 bis 1914. Berlin: Propyläen Verlag.

Landes, David S. 1949. "French Entrepreneurship and Industrial Growth in the Nineteenth Century." *Journal of Economic History* 9: 45–61.

———. 1969. *The Unbound Prometheus: Technological Change and Industrial Development in Western Europe from 1750 to the Present*. Cambridge: Cambridge University Press.

Levine, A. L. 1967. *Industrial Retardation in Britain*. New York: Basic Books.

Mann, Julia De L. 1958. "The Textile Industry: Machinery for Cotton, Flax, Wool, 1760–1850." In *A History of Technology*, edited by Charles Singer, E. J. Holmyard, A. R. H'all, and T. I. Williams. Vol. 4: *The Industrial Revolution 1750 to 1850*. Oxford: Oxford University Press.

Mantoux, Paul. 1961. *The Industrial Revolution in the Eighteenth Century: An Outline of the Beginnings of the Modern Factory System in England*. New York: Harper and Row.

Mathias, Peter. 1969. *The First Industrial Nation: An Economic History of Britain 1700–1914*. New York: Charles Scribner's Sons.

McCloskey, Donald N., and Lars G. Sandberg. 1971. "From Damnation to Redemption: Judgements on the Late Victorian Entrepreneur." *Explorations in Economic History* 9: 89–108.

McNeil, William H. 1982. *The Pursuit of Power: Technology, Armed Force, and Society Since A.D. 1000*. Chicago: University of Chicago Press.

Merton, Robert K. 1970. *Science, Technology and Society in Seventeenth-Century England*. New York: Howard Fertig.

Mitch, David. 1999. "The Role of Education and Skill in the British Industrial Revolution." In *The British Industrial Revolution: An Economic Perspective*, edited by Joel Mokyr. Boulder, CO: Westview Press.

Mokyr, Joel. 1985. "Demand vs. Supply in the Industrial Revolution." In *The British Industrial Revolution: An Economic Perspective*, edited by Joel Mokyr. Boulder, CO: Westview Press.

———. 1999. "Editor's Introduction: The New Economic History and the Industrial Revolution." In *The British Industrial Revolution: An Economic Perspective*, edited by Joel Mokyr. Boulder, CO: Westview Press.

———. 2002. *The Gifts of Athena: Historical Origins of the Knowledge Economy*. Princeton, NJ: Princeton University Press.

Morton, Alan Q., and Jane A. Wess. 1993. *Public and Private Science: The King George III Collection*. Oxford: Oxford University Press.

Multhauf, Robert P. 1967. "Industrial Chemistry in the Nineteenth Century." In *Technology in Western Civilization*, edited by Melvin Kranzberg and Carroll W. Pursell Jr. Vol. 1: *The Emergence of Modern Industrial Society Earliest Times to 1900*. New York: Oxford University Press.

Musson, A. E., and Eric Robinson. 1969. *Science and Technology in the Industrial Revolution*. Manchester: University of Manchester Press.

Nef, J. U. 1957. "Coal Mining and Utilization." In *A History of Technology*, edited by Charles Singer, E. J. Holmyard, A. R. Hall, and T. I. Williams. Vol. 3: *From the Renaissance to the Industrial Revolution 1500–1750*. Oxford: Oxford University Press.

O'Brien, Patrick K. 1985. "Agriculture and the Home Market for English Industry, 1660–1820." *English Historical Review* 100: 773–800.

———. 1991. "The Foundations of European Industrialisation: From the Perspective of the World." *Journal of Historical Sociology* 4.

———. 1993. "Introduction: Modern Conceptions of the Industrial Revolution." In *The Industrial Revolution and British Society: Essays in Honour of Max Hartwell*, edited by P. K. O'Brien and R. Quinault. Cambridge: Cambridge University Press.

———. 1994. Introduction to *The Industrial Revolution in Europe*, edited by P. K. O'Brien. Vol. 1. Oxford: Blackwell.

Ogilvie, Sheilagh C. 1996. "Proto-Industrialization in Germany." In *European Proto-Industrialization*, edited by Sheilagh C. Ogilvie and Markus Cerman. Cambridge: University of Cambridge Press.

Pacey, Arnold. 1980. *The Maze of Ingenuity: Ideas and Idealism in the Development of Technology*. Cambridge, MA: MIT Press.

Parry, J. H. 1966. *The Establishment of European Hegemony, 1415–1715: Trade and Exploration in the Age of the Renaissance.* New York: Harper and Row.

Paulinyi, Akos. 1991. "Die Umwälzung der Technik in der Industriellen Revolution zwischen 1750 und 1840." In *Propyläen Technikgeschichte*, edited by Wolfgang König. Bd. 3: *Mechanisierung und Maschinisierung 1600 bis 1840.* Berlin: Propyläen Verlag.

Pomeranz, Kenneth. 2000. *The Great Divergence: Europe, China, and the Making of the Modern World Economy.* Princeton, NJ: Princeton University Press.

Pursell, Carroll W., Jr. 1997. "Machines and Machine Tools, 1830–1880." In *Technology in Western Civilization*, edited by Melvin Kranzberg and Carroll W. Pursell Jr. Vol. 1: *The Emergence of Modern Industrial Society Earliest Times to 1900.* New York: Oxford University Press.

Rathgen, Bernhard. 1928. *Das Geschütz im Mittelalter.* Berlin: VDI Verlag.

Rosenberg, Nathan. 1972. *Technology and American Economic Growth.* New York: Harper and Row.

Rossi, Paolo. 1970. *Philosophy, Technology and the Arts in the Early Modern Era.* New York: Harper and Row.

———. 2000. *The Birth of Modern Science.* Translated by Cynthia De Nardi Ipsen. Oxford: Blackwell.

Rostow, W. W. 1960. *The Stages of Economic Growth: A Non-Communist Manifesto.* Cambridge, MA: Cambridge University Press.

———. 1963. *The Take-Off into Self-Sustained Growth.* New York: St. Martin's Press.

Schmidtchen, Volker. 1992. "Technik im Übergang vom Mittelalter zur Neuzeit zwischen 1350 und 1600." In *Propyläen Technikgeschichte*, edited by Wolfgang König. Bd. 2: *Metalle und Macht 1000 bis 1600.* Berlin: Propyläen Verlag.

Schubert, H. R. 1966. "The Steel Industry." In *A History of Technology*, edited by Charles Singer, E. J. Holmyard, A. R. Hall, and T. I. Williams. Vol. 5: *The Late Nineteenth Century 1850 to 1900.* Oxford: Oxford University Press.

Shapin, Steven. 1996. *The Scientific Revolution.* Chicago: University of Chicago Press.

Slokar, Johann. 1914. *Geschichte der österreichischen Industrie und ihrer Förderung unter Kaiser Franz I.* Vienna: F. Tempsky.

Smith, Cyril Stanley. 1967a. "Metallurgy: Science and Practice before 1900." In *Technology in Western Civilization*, edited by Melvin Kranzberg and Carroll W. Pursell Jr. Vol. 1: *The Emergence of Modern Industrial Society Earliest Times to 1900.* New York: Oxford University Press.

———. 1967b. "Metallurgy in the Seventeenth and Eighteenth Centuries." In *Technology in Western Civilization*, edited by Melvin Kranzberg and Carroll W. Pursell Jr. Vol. 1: *The Emergence of Modern Industrial Society Earliest Times to 1900*. New York: Oxford University Press.

Smith, Merritt Roe. 1977. *Harper's Ferry and the New Technology: The Challenge of Change*. Ithaca, NY: Cornell University Press.

Stewart, Larry. 1992. *The Rise of Public Science: Rhetoric, Technology, and Natural Philosophy in Newtonian Britain, 1660–1750*. Cambridge: Cambridge University Press.

Thomas, R. P., and D. N. McCloskey. 1981. "Overseas Trade and Empire, 1700–1860." In *The Economic History of Britain Since 1700*, edited by R. C. Floud and D. N. McCloskey. Vol. 1. Cambridge: Cambridge University Press.

Tilly, R. H. 1994. "German Industrialization and Gerschenkronian Backwardness." In *The Industrial Revolution in Europe*. Edited by P. K. O'Brien. Vol. 2. Oxford, Blackwell.

Tracy, J. D., ed. 1990. *The Rise of Merchant Empires*. Cambridge: Cambridge University Press.

———. 1991. *The Political Economy of Merchant Empires*. Cambridge: Cambridge University Press.

Travis, Anthony S. 1993. *The Rainbow Makers: The Origins of the Synthetic Dyestuffs Industry in Western Europe*. Bethlehem, PA: Lehigh University Press.

Troitzsch, Ulrich. 1991. "Technischer Wandel in Staat und Gesellschaft zwischen 1600 und 1750." In *Propylaen Technikgeschichte*, edited by Wolfgang König. Bd. 3: *Mechanisierung und Maschinisierung 1600 bis 1840*. Berlin: Propyläen Verlag.

Tuchman, Barbara W. 1966. *The Proud Tower: A Portrait of the World before the War 1890–1914*. New York: Macmillan.

Tylecote, R. F. 1976. *A History of Metallurgy*. London: Metals Society.

Usher, A. P. 1957. "Machines and Mechanisms." In *A History of Technology*, edited by Charles Singer, E. J. Holmyard, A. R. Hall, and T. I. Williams. Vol. 3: *From the Renaissance to the Industrial Revolution 1500 to 1750*. Oxford: Clarendon Press.

———. 1967. "The Textile Industry, 1750–1830." In *Technology in Western Civilization*, edited by Melvin Kranzberg and Carroll W. Pursell Jr. Vol. 1: *The Emergence of Modern Industrial Society Earliest Times to 1900*. New York: Oxford University Press.

Weber, Wolfhard. "Verkürzung von Zeit und Raum: Techniken ohne Balance zwischen 1840 und 1880." In *Propyläen Technikgeschichte*, edited by Wolfgang König. Berlin: Propylaen Verlag.

Webster, Charles. 1976. *The Great Instauration: Science, Medicine and Reform, 1626–1660*. New York: Holmes and Meier.

Wehler, Hans-Ulrich. 1987–1995. *Deutsche Gesellschaftsgeschichte*, 3 vols. Munich: Verlag C. H. Beck.

Wescott, G. F. 1958. *The British Railway Locomotive*. London: Science Museum.

Wiener, Martin J. 1981. *English Culture and the Decline of the Industrial Spirit, 1850–1980*. Cambridge: Cambridge University Press.

Woloch, Isser. 1982. *Eighteenth-Century Europe: Tradition and Progress, 1715–1789*. New York: W. W. Norton.

Woodbury, Robert S. 1967. "Machines and Tools." In *Technology and Western Civilization*, edited by Melvin Kranzberg and Carroll W. Pursell Jr. Vol. 1: *The Emergence of Modern Industrial Society Earliest Times to 1900*. New York: Oxford University Press.

Wrigley, E. A. 1989. *Continuity, Chance and Change: The Character of the Industrial Revolution in England*. Cambridge: Cambridge University Press.

INDEX

Academy of Experiment (Florence), 23, 61
Agricola (Georg Bauer), 16
agriculture, 6, 18, 39–40, 101
airplanes, 89, 100, 109, 114, 115
aluminum, 100, 108
Apocalypse, 11–12
Archimedes, 4
Arkwright, Richard, 44
artillery, 13–14, 20, 115
assembly lines, 110, 111
automobiles, 97, 98, 100, 108, 110–11, 114

Bacon, Francis, 3–4, 11, 18, 21, 23, 29
Badenese Aniline and Soda Factory, 96, 97
Baeyer, Adolf, 96
battleships, 115
Benz, Karl, 97, 98
Berg, Maxine, x, 38–39, 56–57
Berti, Gasparo, 23
Bessemer, Henry, 89
Biringuccio, Vanuccio, 22
Black Death, 7
Boulton, Matthew, 60
Brahe, Tycho, 24
bronze, 13, 16, 22
Bunsen, R. W., 82, 96

Carnegie Institution, 112
Cartesianism, 30, 61, 62, 65
Cartwright, Edmund, 44
chemical industries, 94–97, 99, 100–101, 104. *See also* sulfuric acid; nitric acid; synthetic dyes

Index

coal, 15, 18, 22, 50, 52, 66, 70, 99, 102
Copernicus, Nicolaus, 24
copper, 13, 22
Cort, Henry, 50
cottage industry. *See* putting-out system
crank and connecting rod, 18
Crompton, Samuel, 44

Daimler, Gottfried, 97
Darby, Abraham, 50
De re metallica, 16
Desaguliers, Jean Theophile, 31–32, 61
Desargues, Gérard, 20
Descartes, Réne, 29–30, 61. *See also* Cartesianism
Dürer, Albrecht, 11, 12

École Polytechnique, 82, 99
educational systems. *See* industrialization, government promotion of in Europe
electrical industries, 92–94, 95, 102, 108, 112–13. *See also* electricity
electricity, 92, 108, 112–13
  scientific discovery and development of, 92–93, 112–13
enclosure, 39
England, 6, 13, 15, 18, 23, 27–60, 66–74, 102–108. *See also* industrialization, in England; science, in the First Industrial Revolution; technological systems, in the First Industrial Revolution
  Glorious Revolution in, 35
  government borrowing in, 35–36

Faraday, Michael, 113
Ferguson, James, 33
Fischer, Emil, 77–79, 112
fly shuttle, 42, 43
France, 13, 61, 64–66, 97–101, 105, 115. *See also* industrialization, in France
Freemasons, 32, 62–63
French Academy of Science, 24

Galilei, Galileo, 21
Galileo. *See* Galilei, Galileo
Germany, 6, 7, 15, 22, 62–64, 72, 77–98, 105, 110, 112, 115. *See also* industrialization, in Germany
Gerschenkron, Alexander, 83
Gilchrist, Sidney, 90
Gilchrist Thomas, Sidney, 90
Gille, Bertrand, x, 16–18, 71, 73–74, 116
Graebe, Carl, 96
Great Instauration, The, 4, 5
Griffith, William, 33
Guericke, Otto von, 23
guilds, 63, 64, 79–80
Gutenberg, Johannes, 6

Haber, Fritz, 97
Hargreaves, James, 44
Hero of Alexandria, 23
Hèroult, Paul L. T., 100
Hoffmann, August Wilhelm, 94
Horsley, John, 32
Huygens, Christian, 24

industrialization, 22. *See also* industrial takeoff; science, in the First Industrial Revolution;

science and technology, interaction of
  economic origins of, 34–41
  in England, 25, 34–60, 66–74, 102–108
  in France, 61–62, 64–66, 73, 75, 97–101, 105
  in Germany, 63–64, 73, 75, 77–97, 105, 110–12, 114
  government promotion of in Europe, 79–82
  recent interpretations (First Industrial Revolution), 54–60
  in the USA, 75, 77–79, 108–12
industrial takeoff, 54
  in England, 54
  in France, 99
  in Germany, 83–84
interchangeable parts, 66, 89, 109–10, 114
internal combustion engines, 89, 97
iron, 13–15, 16, 18, 36, 48–54, 64, 66, 69, 70, 71, 73, 74, 83, 84, 85, 89, 99, 113. *See also* potting-and-stamping process; puddling process; steel
Italian Wars (1494–1559), 11, 13, 20
Italy, 6, 7
itinerant lecturers, 27–28, 33

Johnson, Samuel, 23–24

Kaiser Wilhelm Society for the Advancement of the Sciences, 77, 78
Kay, John, 42, 43
Kepler, Johannes, 24, 29
Knietsch, Rudolf, 97

Leblanc, Nicolas, 100
Liebig, Justus, 82, 94
lightbulbs, 109

machine guns, 109
machine tools, 18, 65–66, 69–70, 71, 84, 97, 106, 109–10, 114. *See also* interchangeable parts
Martin, Pierre, 90
military technology. *See* technology and war
millenarianism, 11, 19
mining, 15–16, 18, 63, 70, 74
Mokyr, Joel, x, 58–59

navigation, 4–5
New Atlantis, The, 1–3, 11, 23
Newcomen, Thomas. *See* steam engines, of Thomas Newcomen
Newton, Sir Isaac, 25, 29–30. *See also* Newtonianism
Newtonianism, 29–32, 61–62, 81
nitric acid, 97, 114, 116

Oersted, Christian, 113
Otto, Nicolaus, 97

Papin, Denis, 23
Parsons, Charles, 102
Perkin, William Henry, 94
petroleum, 108
Plato, 3
population growth, 6, 34, 37, 62, 63, 101
Porta, Giambattista della, 23
potting-and-stamping process, 50, 52
power looms, 44, 52, 68, 84
Principia, 25

printing, 6–7, 8, 16
proto-industrialization, 38–39, 63
puddling process, 50, 52, 53
putting-out system, 38, 41, 52, 63, 64

radios, 114, 116
railroads, 70–71, 72, 73, 81, 85, 99
Ramsay, David, 22
Rathenau, Emil, 94
Regiomontanus, 20
Rockefeller Institute for Medical Research, 112
Roebuck, John, 46, 60
Rostow, Walter W., 54, 83–84. *See also* industrial takeoff
Rotheram, Caleb, 33
Royal Society of London, 23, 25, 31
Russia, 115

science. *See also* Cartesianism; Newtonianism
   Aristotelian school of, 3, 20, 24, 61–62
   in the early modern period, 3–4, 19–25
   in the First Industrial Revolution, 27–34, 61–65, 81–82
   magical or organicist (Platonic) school of, 3, 22, 24, 61
   mechanical (Archimedean) school of, 3, 20, 24, 30, 62
   in the Second Industrial Revolution, 81–83, 92, 94, 96–97, 107–108, 110–11, 112–14
science and technology, interaction of, 19–25, 27–34, 61–65, 77–79, 81–83, 92, 94, 96–97, 99, 107–108, 110–11, 112–14
Shakespeare, William, 11

shipbuilding, 4–5, 15
Siemens, William, 90
Smeaton, John. *See* steam engines, of John Smeaton
Smith, Adam, 37
soda, 100–101, 104; Leblanc process for, 100, 101, 104; Solvay process for, 100–101, 104
Solvay, Ernest, 100
Spalding Gentleman's Society, 31
spinning or water frame, 44, 45
spinning jenny, 44, 45, 68
spinning mule, 44, 66, 67, 84
spinning wheels, 7, 9, 18
Spinoza, Benedict de, 30
steam engines, 22–23
   compound, 70
   development of first, 22–23
   high-pressure, 70
   of Thomas Newcomen, 23, 46–48
   of John Smeaton, 48, 60
   turbine, 89, 102–103, 109, 113
   of James Watt, 48, 49, 60, 70, 104
steel, 18
   basic, 90, 100
   Bessemer, 89–90, 91, 100, 104, 105
   electric arc, 100
   open hearth, 90, 100, 104, 105, 110
   puddled, 89
Stephenson, George, 71
Stevin, Simon, 20
sulfuric acid
   contact process for, 69, 76, 90, 114, 115
   lead-chamber process for, 46, 68–69, 73, 90, 114
synthetic dyes, 89, 94, 96, 112, 114

Tartaglia, Niccolo, 20
technological systems, x. *See also* Gille, Bertrand
  contemporary, 117
  in the early modern period, 16, 18–19
  in the First Industrial Revolution, 41–42, 52, 54, 66–75, 79, 85–89
  in the Second Industrial Revolution, 85–116
technology and war, 11–18, 62, 65–66, 109, 115–16, 117. *See also* artillery; battleships; machine guns
telephones, 89, 109, 114, 116
Tempest, The, 11
Tennant, Charles, 68
textiles, 16, 36, 37, 38, 41–46, 64, 66–68, 74, 84, 99, 102. *See also* proto-industrialization; putting-out system

Thirty Years' War (1618–1648), 11, 18, 62

United States of America, 75, 108–12. *See also* industrialization, in the USA
  colonial beginnings of, 34–35

Vinci, Leonardo da, 19, 20

Walker, Adam, 27–28, 33
Watt, James. *See* steam engines, of James Watt
Winckler, Clemens, 97
windmills, 7, 10, 18, 71
Wöhler, Friedrich, 82
World War One, 115–16

x-ray diffraction, 113